ROADS AND DEVELOPMENT IN SOUTHWEST BOUGAINVILLE

Plate 1 Sikoreva Hill on the new trans-island road. The Jaba River Pump Station is visible on the land between the valley of the Kawerong River (top) which has been heavily polluted by mine waste, and the clear Jaba River which is crossed by a low-level concrete crossing.

New Guinea Research Bulletin No. 62

ROADS AND DEVELOPMENT IN SOUTHWEST BOUGAINVILLE

MARION W. WARD

New Guinea Research Unit
The Australian National University
Port Moresby and Canberra
1975

Printed and manufactured in Australia by
The Australian University

National Library of Australia
card no. and ISBN 0 85818 028 6

Contents

Figures

Plates

Preface

This study was undertaken in 1972 as part of a program of research on Bougainville Island financed through the Pacific Islands Research Corporation, University of Hawaii.

My aim was twofold: to examine the initial impact of the trans-island road, and to obtain base-line data which could be utilized in a possible re-study to be carried out after the road has been open for some years and patterns of movement have stabilized into a new form.

A relatively short period of field work was undertaken in Port Moresby and Bougainville during October 1972. Analysis of the data and writing was carried out in Canberra between November 1972 and February 1974. A draft report was circulated widely during 1974 and the comments and suggestions made by readers incorporated in this Bulletin. I am grateful for the assistance received in this respect, but the responsibility for any errors which remain is mine.

The original draft used non-metric weights and measures. For the most part these have now been converted to the metric system, but the units used in original sources have been indicated where necessary. The freight costs in Table 3.21 and the information in Tables 7.5-7.10 and the relevant text, together with the calculations of benefit-cost ratios given in Chapter 8 and Appendix V have not been converted. Since benefit-cost ratios and internal rates of return, which are an important part of the study, are themselves ratios conversion was not necessary.

The spelling of village names presents particular problems in a study such as this. After much consideration a decision was made to standardize to those spellings given in the Village Directory, 1973, despite its obvious infelicities. Readers familiar with the area will recognise both the errors and the difficulties.

Marion W. Ward

New Guinea Research Unit
A.N.U., Canberra
November 1974.

Acknowledgments

I wish to express my thanks to all those who assisted in this study. These include, above all, the people of South-west Bougainville who made us feel welcome in the area, expressed interest in the study, assisted us in many ways but particularly with transport and accommodation, helped us across rivers, answered our questions, and tolerated our Pidgin. I hope that this study will contribute to wise planning of the development of their homeland.

With the achievement of self-government in December 1972 many of the officials and others who helped us have since moved to other places and positions, and the names of some government departments and other organizations have changed. Those mentioned below apply to October 1972.

In Kieta thanks are due to the then District Commissioner and his staff, the Superintendent of Police and staff, staff of the Rural Development Division, Department of Labour, Development Bank, Business Development and Co-operatives Divisions, Department of Trade and Industry; Mr Francis Seeto; Mr B.F. Darcey; the manager, Bougainville Air Services.

In Southwest Bougainville to Mr G. Schweinfurth, Patrol Officer, Boku; Mr M. Lang, A.D.C. Buin; Mr K. Fairweather, Co-operatives Division; Mr P. Wilson and Mr B. French, DASF Konga and Buin; Mr N. Phipps; Father W. Fingleton; Mr R. Blanchfield; Mrs Kerr; Mr Fordham; officers of the Siwai, Bana and Buin Producers' Co-operative Societies; officers of the Bana, Siwai and Buin Local Government Councils.

At Bougainville Copper Proprietary, Panguna, to Mr K. McKenzie, Mr C. Newman, Mr B. Crawley, Mr E. Djokovic and other staff.

At Port Moresby to Mr I. Burnet, Mr D. Smith and other officials of the Department of Transport; Mr H. Jackman and members of the staffs of the Co-operatives Division and Trade Store Sections of the Department of Trade and Industry; the Town Planning Section, Department of Lands, Surveys and Mines; the then Department of District Administration library; the Department of Public Works; the then Department of Civil Aviation; the Papua New Guinea Development Bank; and to the New Guinea Research Unit, The Australian National University, particularly Dr R. May, Drs D. and J. Mitchell, and Mr J. Toner.

In Sydney to Mr G. McDonell for advice on the determination of benefit-cost ratios for the roads.

In Canberra to Mr J. McAlpine, C.S.I.R.O.; Mr E.K. Fisk, Dr R.M. Sundrum, Dr M. Treadgold and other members of the staff of the Department of Economics, Research School of Pacific Studies; Professor D. Freeman, Department of Anthropology, for the use of office space; Dr D. Laycock of the Department of Linguistics.

A special debt of gratitude is owed to Mrs Amirah Inglis who acted as research assistant; to Dr Diana Howlett and Mrs Nancy Clark of the Department of Human Geography, A.N.U., for assistance in preparing the first draft for publication in my absence; to Mr Hans Gunther, Cartographic Unit, Department of Human Geography, A.N.U., who drew the maps; Mrs E. Cochrane and Mrs D. Mitchell who helped earlier and final drafts of the manuscript; Mrs S. Purchase for editorial assistance; Mrs M. Dudley; and to my husband, Professor R.G. Ward, for professional advice and personal encouragement.

Marion W. Ward

Chapter 1

Introduction

In the latter months of 1972 about 22 km of road were constructed from Moratona on the lower course of the Pangala River to the Jaba River Pump Station in west central Bougainville Island. The Pump Station is part of the Bougainville Copper Proprietary Limited (BCP) mining operation, and from it a good road connects to Arawa and Kieta on the east coast. Moratona was the northernmost point on the existing road network of Southwest Bougainville, hence the construction of the new road gave all of Southwest Bougainville road access to the urban centre and overseas port of Kieta. This area had a population of about 32,000 in 1972, and is an important cocoa producing area.

This situation offered conditions for a study of the potential effects of new road construction in Papua New Guinea, but with relevance also to the wider context of the developing world. By observing the position shortly after the road was opened base-line data were collected which can be compared with the situation as it may develop in some years' time. This study presents the results of field observations made during October 1972, at which time the road had been in use for about two months. It will also contribute to the increasing volume of information about Bougainville in the 1970s, being provided by the work of Ogan (1972), Griffin (1972), D. Mitchell (1971), J. Mitchell (in press), Couper and Cotter (1972), Moulik (1974), and Bedford and Mamak (in progress).

Transport development in Papua New Guinea

The past decade in Papua New Guinea has been a period of rapid development of modern land transport facilities, which, apart from the often ephemeral constructions of World War II, had previously been minimal. The earliest indigenous pattern of communications was a network of foot-tracks and canoe routes, some of which supported extensive systems of trade and exchange. In the earlier decades of European contact emphasis was placed on the creation of ports, supplemented by local road systems of limited extent. These were often no more than tracks suitable for walking, horse-riding, bicycles, or occasionally wheeled vehicles. One exception was the relatively good vehicular road built in German times fringing the east coast of New Ireland. On the mainland the value of air transport in mountainous unroaded country was demonstrated by the development of the Wau-Bulolo goldfields during the 1930s. World War II showed even more clearly that airstrips could be readily built and that air transport could provide a network of needed communications. Both the Japanese and the American and Australian forces made use of existing and previously undeveloped natural harbours, and built roads hastily in locations determined by military necessity. Many of the wartime facilities were allowed to decay after the war.

The 1950s saw a surge of developmental airstrip building particularly on the mainland, and local road networks began to be constructed to service major airfields in the interior. Not until the opening of the Highlands Highway linking Lae with Goroka and Mt Hagen in the mid-1960s was any extensive road connection completed. Since then a number of major road projects have been undertaken pushing out from the urban centres of Port Moresby, Lae, Goroka, Mt Hagen, Madang, Wewak and Rabaul. These have been supplemented by local feeder road construction in many parts of the country. In another one or two decades it is likely that a highway network linking most of the major urban centres on the mainland, and encircling the major islands of New Britain, New Ireland and Bougainville will be completed.

At present there is in Papua New Guinea a heavy emphasis on road construction and upgrading, because road transport is more flexible and convenient. Major port and airfield developments, while still continuing where necessary to accommodate ever-increasing volumes of movement, are no longer relied on as a means of opening up areas, except in extremely isolated locations. Railways have only ever been used in special circumstances such as to bring ore from a mine to the coast, and then have been narrow-gauge and short in length. It is unlikely that major general purpose railways will ever be built in Papua New Guinea.

Transport development in Bougainville

The development of transport facilities on Bougainville Island fits this general pattern. In pre-European times, as

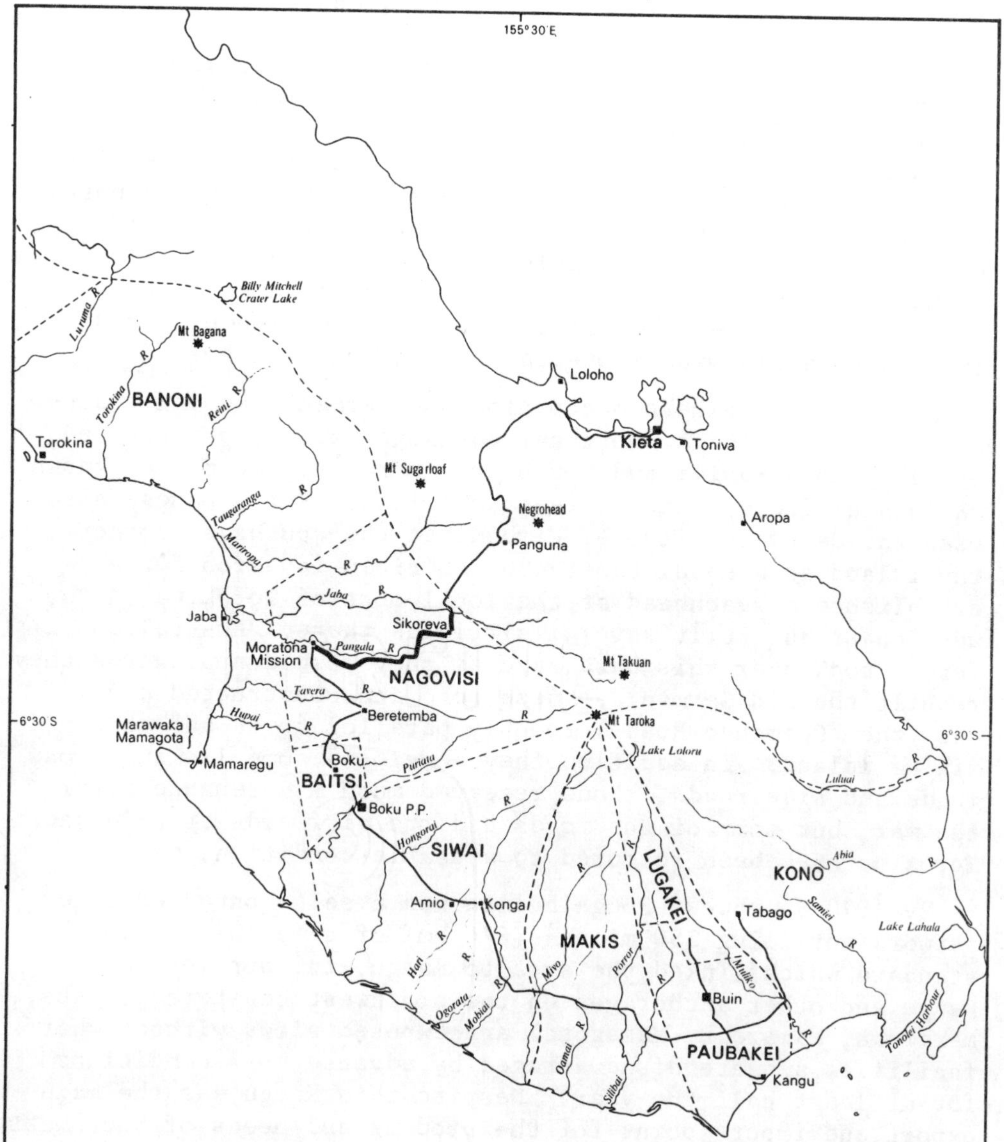

Figure 1.1 Location of the trans-island road in Southwest Bougainville

still today, villages were presumably linked by foot-tracks, and canoes were used around the coasts, and to reach offshore islands and more distant island groups. The first mission stations (after 1902), the first German administrative centre (Kieta, established 1905), the first Australian administrative centres of Sohano (at Buka Passage) and Kangu/Kihili (on the south coast), established before World War II, and the first European-owned plantations (on the east, northeast and north-west coasts and dating from before World War I) were all coastal. In Southwest Bougainville the German administration required the construction and maintenance of a main 'road' around the southwest quadrant of the island, and of 'roads' (more accurately wide foot-tracks) linking villages.

During the Japanese occupation the German road and tracks were utilised, and a short narrow-gauge railway constructed between Kangu/Kahili and Turiboiro (Kara) airfield (now known as Buin airfield) some fifteen km inland. The Japanese also used the deepwater Tonolei Harbour at the southeast corner of the island as a naval base. The American invasion force established a beachhead at the local port of Torokina on the west coast and built several airfields there. Australian forces took over this base, and as they pushed southwards they rebuilt the old German/Japanese road and constructed a new one (the 'Commando Road') roughly parallel to it and about six km inland. In addition they developed some linking cross roads and side roads. None received much maintenance after the war, but some of the roads and the airfields at Buin and Torokina have been restored to a usable condition.

By 1970 Southwest Bougainville had a self-contained road network totalling 390 miles[1] (624 km) of a medium to low standard which linked the area to Kangu, the port of Buin. Kangu and other anchorages on the southwest coast (e.g. Jaba, Marawaka, Mamaregu, Mamagota) are exposed sites without wharf facilities and are disadvantaged by adverse surf conditions for at least half the year. Despite this Kangu was the main export and import point for the produce and needs of Southwest Bougainville until completion of the new road.

Pressure for further road construction at both trunk and feeder levels is now intense in Bougainville. People

---

[1] Of this 150 miles (240 km) were described as suitable for light and medium traffic; and 240 miles (384 km) suitable for four-wheel-drive vehicles only. There were also 500 miles (800 km) of bridle paths. (Bougainville District Annual Report, 1969-70:Appendix 10.)

increasingly want to be able to ship their produce directly to overseas ports at the lower cost made possible by direct road transport. They also want the personal convenience of direct access to the growing urban centres of Kieta-Arawa-Panguna. The demand for new or improved road access is not likely to decline until a trunk road system covering the whole island is completed, and until most villages are served by reliable feeder roads.

Studies on the effects of new roads

Concurrently with the increased pace of development of transport facilities in Papua New Guinea and elsewhere a demand has arisen for in-depth studies to discover the effects of transport projects on the people and economies of the area affected. Governments and other fund-providing agencies have been interested, for planning purposes, in benefit-cost appraisals of a variety of projects.

Some of the more notable of such studies have been carried out in countries in Latin America by the Brookings Institution (Wilson *et al*., 1966). In Uganda, Smith examined the relationships between increase in road mileage, population and cotton production in three regions (Smith, 1959). In Sabah, Bonney studied the relationships between road-building and economic and social development (Bonney, 1964), and was followed by Sabag with a case study of the Tenom-Tambunan valley road link (Sabag, 1971). Several studies of the economic effects of new highways and village roads have been carried out by Jones and others in north and northeast Thailand (Jones, 1964; Jones and Orr, 1966; Patanapanich, 1964; [Thailand] National Statistics Office, 1966; United States Operations Mission, 1966; Kasiraksa, 1963; Ngamsit, 1966; Trikamol, 1966).

In Papua New Guinea the first case study of the effects of new road construction was made of the Port Moresby-Rigo road (Ward, 1970), followed by a study of the effects of feeder roads in the Western Highlands District (Southern, 1970a; 1970b; 1971), and a study of the impact of roads on the monetary activities of subsistence economies in the Okapa and the Gumine regions of the Eastern Highlands District (Bouchard, 1972; 1973). A national survey of transport needs in Papua New Guinea which contains detailed benefit-cost analysis of specific major roads and other transport projects, including consideration of the then transport needs of Southwest Bougainville, was carried out in 1968 by a team under contract to the United Nations Development Bank and the International

Bank for Reconstruction and Development (Halcrow, 1969). In addition benefit-cost studies are made internally in the Papua New Guinea Department of Transport for all major transport projects under consideration. This department has also carried out a number of unpublished traffic and in some cases freight surveys of selected major roads (Highlands Highway Freight Survey, 1968, 1969 and 1970). There have also been engineering feasibility studies for projected major and minor road construction in many parts of Papua New Guinea.

A study with emphasis on freight forwarding through the port of Lae and along the Highlands Highway was made by Rimmer (1972), and two other studies concerned with transport in the Sepik and the Highlands have been published (Philpott, 1972 and Southern, 1973).

All of these contribute to knowledge about the effects of new or upgraded roads and improved transport systems in environments which are often severe in terms of constraints of landforms and/or climate, and in which the economies of the people are largely subsistence but with a potential for increased income through cash-cropping. These studies have relevance for many poorer countries. The experience they report should have value for planners concerned with the allocation of limited resources to the task of providing access to the wider world.

The present study describes Southwest Bougainville and its existing patterns of transport and freight movements; its economy and the potential effects of the new road on this and on the relations of the area with the urban complex of Kieta-Arawa-Panguna; gives benefit-cost analyses for the new trans-island road, other sections of the trunk road network in southern Bougainville, and for feeder road circuits and branches in the area; and discusses supplementary development inputs needed to ensure that the greatest possible benefits to the people of the area are achieved from investment in roads.

Chapter 2

The physical and human environment

The area with which this study is principally concerned is the southern half of Bougainville Island, in particular the Buin subdistrict (Fig.2.1). In administrative terms this comprises the Banoni, Nagovisi, Baitsi, Siwai, Makis, Lugakei, Paubakei and Kono census divisions of Buin subdistrict. The study is concerned to a minor extent with the six southernmost census divisions of the Kieta subdistrict, east of the ranges. Southern Bougainville is about 64 km across at its widest (in a southwest-northeast line) and measures about 80 km from Mt Bagana in the northern Banoni census division southeast to the south coast. The area under consideration (Buin subdistrict) is about 3,781 km^2.

Geology and landforms[1]

The southern half of Bougainville Island is dominated by the andesitic volcanic complex of Mts Taroka and Takuan rising to 2,300 m. In the north of Banoni census division Mt Bagana, an active ash-producing volcano, rises to 1,700 m. The southeast-northwest trending Crown Prince Range reaches 1,200-1,500 m and links the two volcanic complexes. In Kono census division of the southeast the lower Deuro Range rises to about 600 m, continuing the general northwest-southeast trending line of the island, and effectively dividing Buin subdistrict from Kieta subdistrict at the southeast corner of the island. The generalised geology of the area is shown in Fig.2.2.

Falling away from the mountain core to the south and west is an area of well-drained sloping volcanic-alluvial fans and plains. They form a zone about 24-32 km wide which is crossed by many rivers radiating from the mountain core. Inland these streams occupy deeply incised ravines, but within about 16 km of the coast they emerge to form braided channels across the

[1] Material in this and following sections of this chapter is drawn primarily from the fuller accounts in CSIRO, 1967.

Figure 2.1 Administrative divisions

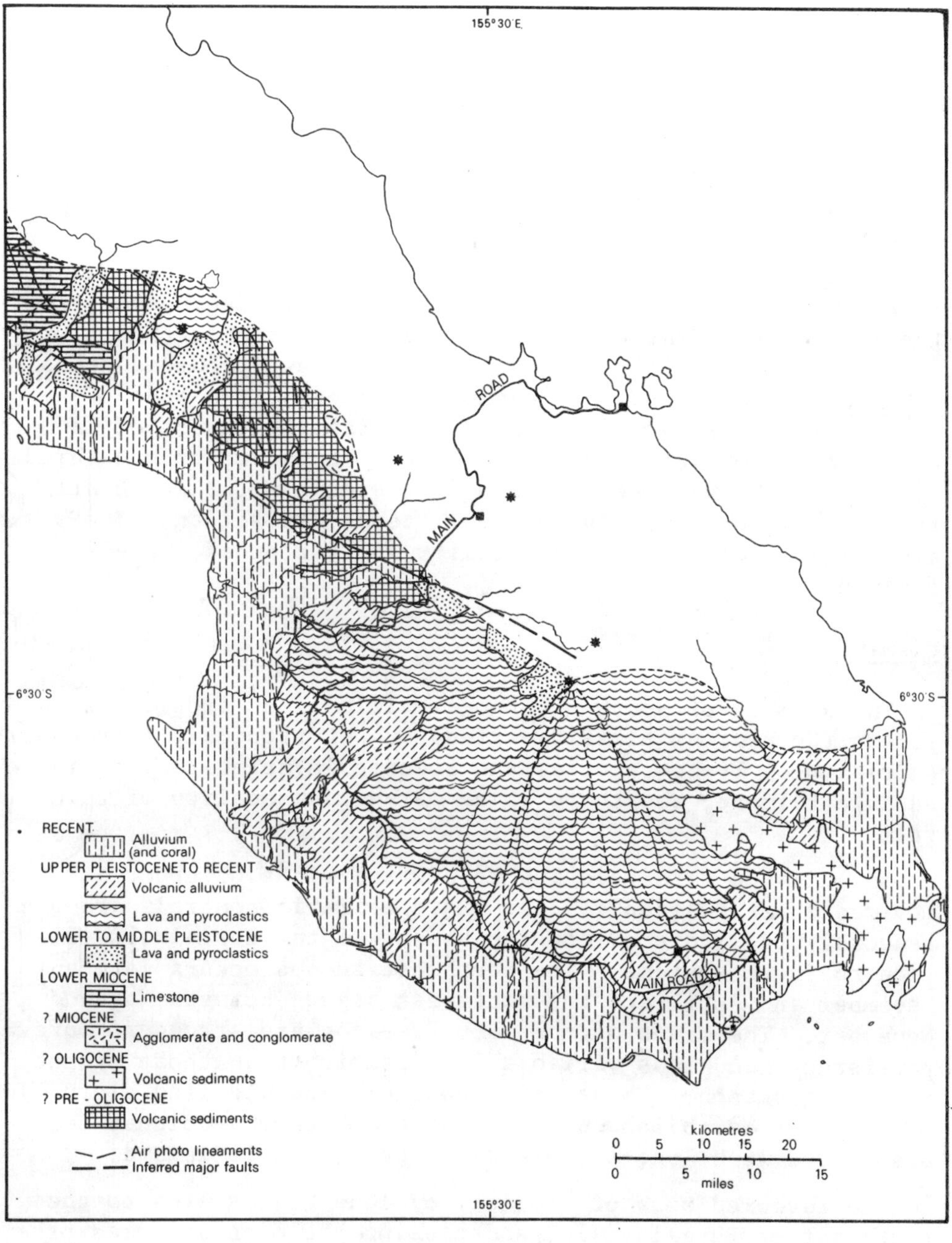

Figure 2.2 Generalised geology

floodplain. In both situations they can be serious barriers to movement, particularly in times of flood. Near the coast, especially in the north of the area, large areas of swamp are common. East of the central ranges in Kieta subdistrict, mountainous and hilly country comes nearer to the coast and the area of coastal plain is limited, especially near Kieta.

South Bougainville is for the most part covered with a mantle of volcanic ash (Fig.2.3) which around the Mts Taroka-Takuan complex is of Early Recent or Pleistocene age, and is now immaturely weathered. This ash layer is more than a metre thick in central Bougainville, but lessens in depth away from the mountains. Near Mt Bagana there is an area covered by modern ash, and this volcano is known to have produced *nuées ardentes* in 1950 and 1952. Around Mt Bagana is a zone of Recent lapillitic ash, that is ash containing layers of lava or pumiceous pellets, and these layers may be up to one-third of a metre thick. The only ash-free areas in the south are the coastal zones of fluvial or littoral deposition. The area is one of tectonic instability and earthquakes are frequent.

Climate

The climate of Southwest Bougainville is of tropical rain forest (Köppen, Af classification) or wet tropical (Thornthwaite, AA'r) type. Its equatorial and oceanic location results in an equable climate, and there is no marked seasonality of rainfall.

No reliable data relating to surface winds or upper air movements are available. The major climatic controls are the seasonal wind systems, and the aspect of the terrain in relation to these systems. The northwest season occurs between December and April and the southeast season between May and November. The winds of the latter season are stronger, more persistent, and less variable in direction than those of the northwest season. These major controls are modified diurnally through the establishment of an orographic-convectional cell associated with land-sea breezes over the whole island.

The favoured zone of tropical cyclone tracks lies to the southeast of Bougainville, and passing cyclones at times produce short periods of heavy rains and strong winds. In May 1972 cyclone Ida passed over Bougainville producing exceptionally heavy rainfall which caused severe flooding by rivers in the southwest.

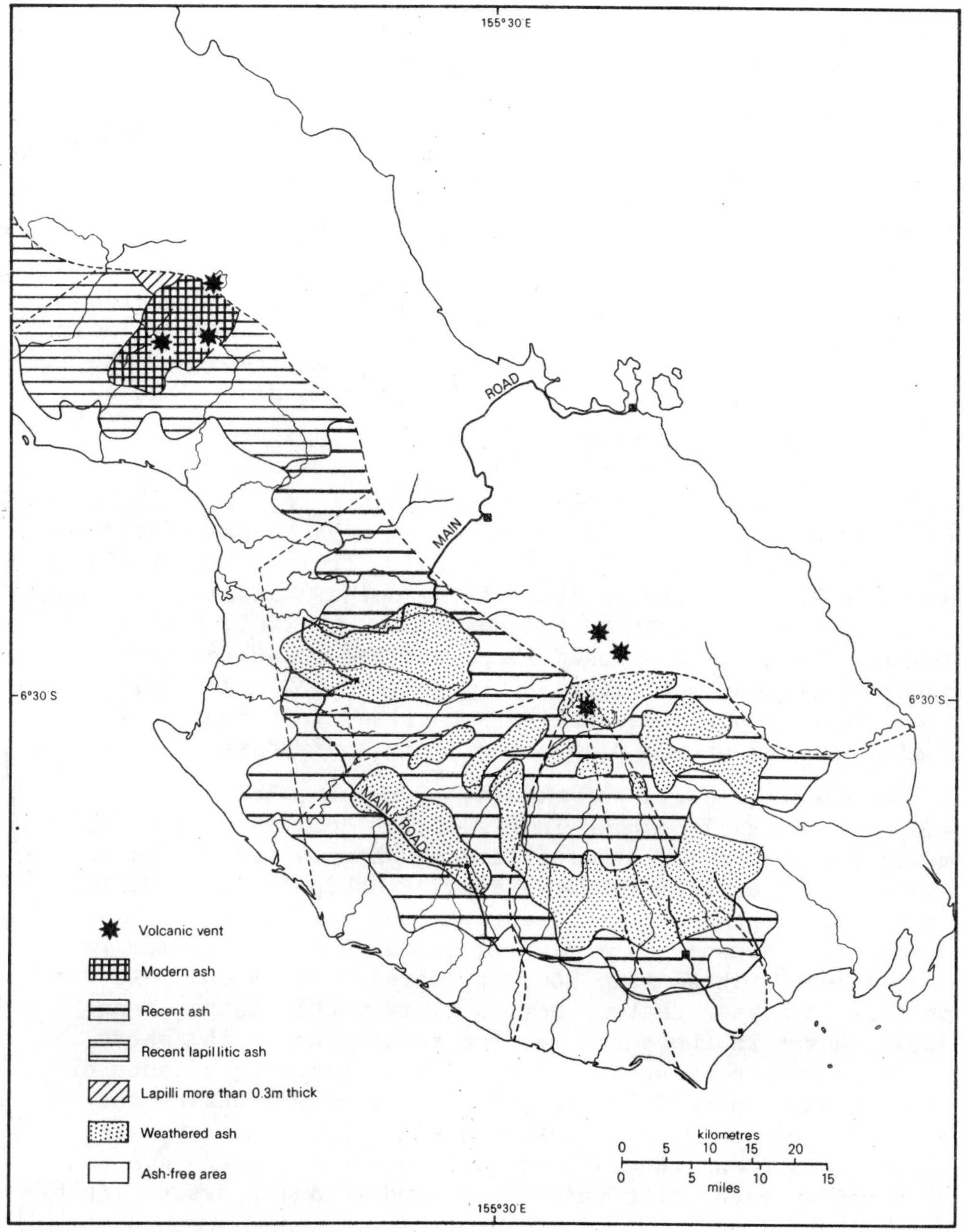

Figure 2.3 Ash cover

Climatic records for the area are intermittent and sketchy but the following details are culled from various sources to give some indication of climate elements.

Rainfall characteristics. Rainfall records are available in some detail only for Buin, Boku and Mokopolo (Moratona) in the southwest area, and are shown in Table 2.1. Panguna, though outside the area, is included as representative of mountain conditions.

Records kept for Buin by the Department of Agriculture, Stock and Fisheries (DASF) since the CSIRO study show wide variation from the CSIRO maximum and minimum figures, namely, in 1969 Buin recorded 536 cm and in 1971 only 246 cm. The CSIRO figures consolidated with DASF figures give Buin an annual average of 376 cm. By the end of August 1972 Buin had already had 373 cm of rain.[1]

An attempt to map rainfall isohyets for Bougainville Island is made in Fig. 2.4. This is based on maps in Brookfield and Hart (1966:Map 5) and CSIRO (1967:64). It is apparent that northern Bougainville is drier than southern and that the west coast is much wetter than the east. It is likely that mean average rainfall on exposed slopes of the mountains in the interior reaches well over 500 cm per annum. Heavy falls in the mountains often cause rivers to rise, interrupting road communications on the plains for periods of hours or days.

Annual distribution of rainfall for Buin, Boku, Mokopolo and Panguna is also shown in Fig. 2.4. Buin shows a definite maximum during the southeast season, whereas Boku shows this and a second maximum in March-April. Mokopolo shows maxima in May and September with drier periods in June-August and November. Panguna has a maximum in March and a minimum in June-August. These are, however, merely wetter and less wet periods in a year that is wet in every month; there is no 'dry' season in the south in contrast to the north, where, at Sohano, Soraken, Kunua and Baniu, the sheltering effects of the central mountains do create a drier season during the southeast period. The total rainfall for Panguna appears rather less than expected, probably because records are taken at a valley floor site which is ringed on all sides by higher mountains.

---

[1] The data in this paragraph are drawn from Transport Dept file no.68/73, Part 4, Bougainville District Central, letter dated 20 September 1972 from D.C. Bougainville to Secretary, Dept of Transport.

Table 2.1

Rainfall characteristics for Buin and Boku

Station		Height	No. of complete years	J	F	M	A	M	J	J	A	S	O	N	D	Annual	Highest annual	Lowest annual
Buin (A)	Mean rainfall (cm)	12m	5	23	21	35	23	16	30	51	29	33	23	32	22	338	391	301
Buin (A)	Mean no. of rain days		5	16	17	18	17	17	17	23	19	17	18	20	17	216		
Buin (B)	Mean rainfall (cm)		12 (to 1963)	21	21	27	23	28	31	52	42	34	33	26	20	358	-	-
Boku (A)	Mean rainfall (cm)	61m	9	40	27	48	46	29	42	42	56	45	43	46	41	505	607	466
Boku (A)	Mean no. of rain days		9	22	14	18	21	16	20	22	22	22	20	17	18	232		
Mokopolo (Moratona) (C)	Mean rainfall (cm)	n.a. (lowland)	2	32	40	38	36	54	24	30	21	42	27	13	24	381	n.a.	n.a.
Mokopolo (moratona) (C)	Mean no. of rain days		2	22	20	24	23	28	25	26	20	21	20	18	21	268		
Panguna (C)	Mean rainfall (cm)	n.a. (mountain)	7	51	48	58	45	33	22	37	26	33	33	37	40	463	n.a.	n.a.
Panguna (C)	Mean no. of rain days		7	27	23	26	25	23	21	29	23	23	26	26	27	299		

Sources: (A) CSIRO, 1967:63.
(B) Brookfield and Hart, 1966:Table I, n.p. (originally in points).
(C) BCP Panguna (originally in points).

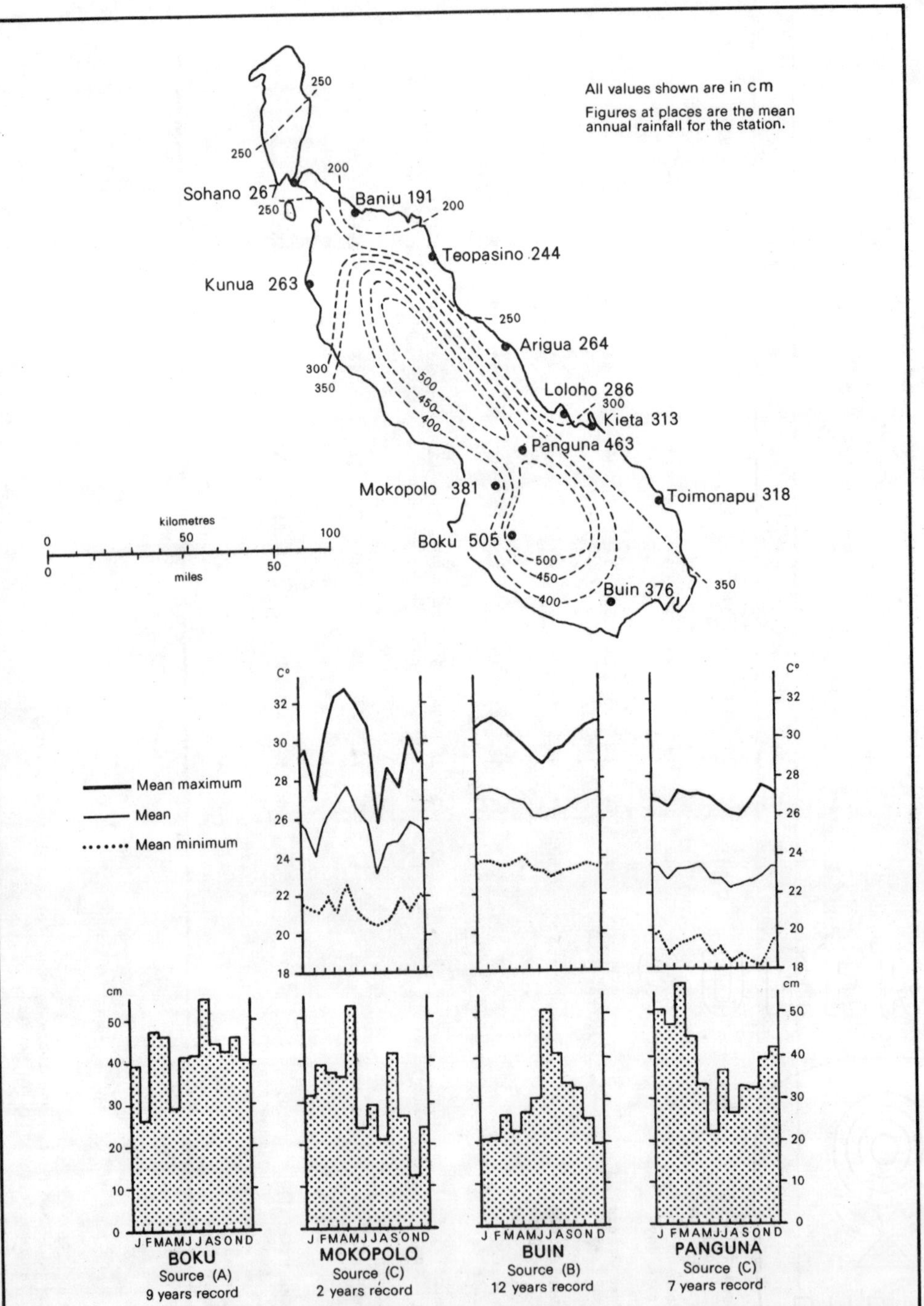

Figure 2.4 Rainfall and temperature

Temperature characteristics. Temperatures in Southwest Bougainville are generally uniform throughout the year. The data for Buin show a mean temperature fluctuating only between 26°C in June and July and 27.4°C in February (Fig.2.4). Mean monthly maxima rise only to 30.9°C and mean monthly minima fall only to 22.8°C. The diurnal temperature range is essentially constant at 7°C, ranging from 23.3°C to 30.5°C. Buin, and the Southwest generally, is cooler than it would otherwise be because it is affected by cold descending mountain air masses (katabatic winds). Mokopolo (on three years of records) shows much greater mean temperature range (23°C to 27.5°C), with the highest mean monthly maximum rising to 32.7°C and the lowest mean monthly minimum falling to 20.4°C. This greater range may result from lower cloudiness at Mokopolo. At Panguna temperatures are again more even, though lower than at Buin. The mean monthly temperatures range from 22.1°C to 23.3°C, while the highest mean monthly maximum is 27.3°C, and the lowest mean monthly minimum is 18.1°C. It may be assumed that normal lapse rate conditions apply, namely 1.7°C drop in mean temperature for every 300 m ascent. There is no frost risk at any elevation.

Relative humidity and insolation. Relative humidity ranges between 79 per cent and 86 per cent and there is little seasonal or diurnal variation. Under these conditions of rainfall and humidity moisture deficits are unlikely to restrict plant growth at any time of the year in Southwest Bougainville.

Cloudiness is higher in Southwest Bougainville than in the east and north. An estimate of 1,304 hours of sunshine for Buin compared with an estimate of 2,055 hours at Aropa is given for the period September 1968 to September 1969 in Sumbak (1970:4) who thinks that this may be the main reason for the poor growth of coconuts in Southwest Bougainville. A record of 1,841 hours of bright sunshine was obtained at Pomaletti village, Nagovisi Census Division, between July 1969 and June 1970 inclusive (pers. comm., D. Mitchell, 31 May 1974).

Soils

The soils of Southern Bougainville have been formed mainly by climatic action on parent materials of volcanic origin. Their distribution in the area under study is shown in Fig.2.5. The main types occurring in the area are:

Swamp peats, consisting of surface peats with minor submerged peats, neutral peaty clays, and brown mottled sands;

Swamp and littoral soils, consisting of neutral peaty sands and brown sands with minor white sands;

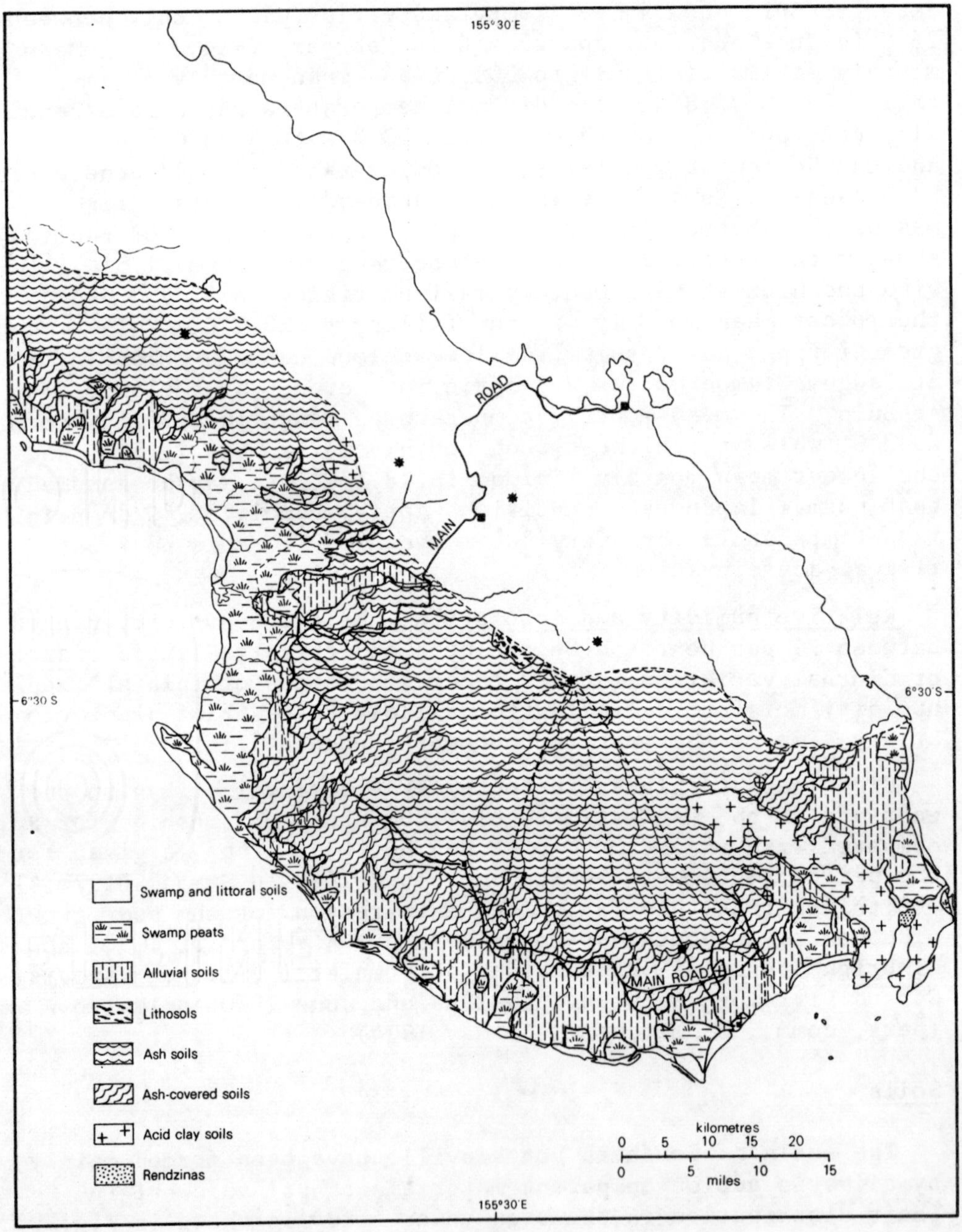

Figure 2.5 Soils

Lithosols, consisting of lithosols with minor brown loams;

Alluvial soils, consisting of brown mottled sands;

Ash soils, consisting of (a) grey fine sands (stony phase), (b) brown loams with lapillitic horizons with minor brown loams locally, (c) brown loams with lapillitic horizons and brown loams, (d) brown loams with minor lithosols locally, (e) brown loams with minor brown loams with an ash pan, and (f) brown loams with an ash pan with minor brown loams and brown mottled sands (stony phase);

Ash-covered soils, consisting of brown loams over sands;

Acid clay soils, consisting of brown friable clays with minor lithosols and brown loams over brown friable clays.

Generally, apart from large areas of swamp soils in the Banoni census division and smaller areas of swamp soils along the coast of Southwest and south Bougainville, the acid clay soils of the Deuro Range, and small areas of lithosols on the mountains, the plains and fans of the Southwest are mostly covered in brown loams with increasing ash cover or ash pans as one moves inland towards the mountains.

In spite of the recent ash cover with its high organic topsoils in the Siwai and Buin areas, available analyses indicate that these soils have low fertility (Scott, CSIRO, 1967:171). The suggestion has been made that trace element deficiencies may be responsible for poor growth of coconut palms in the Southwest (pers. comm., French and Wilson, DASF, Konga and Buin respectively, October 1972).

Vegetation

The natural vegetation cover of the better drained lowlands is tall tropical mixed-species rain forest in which *Vitex* spp. and *Pometia* spp. are characteristic, and *Artocarpus altilis* and *Albizia falcata* are characteristic of secondary forest. *Terminalia-Campnosperma* forest is characteristic of swamp areas, and a coastal fringe of *Casuarina equisetifolia* forest occurs on the outermost beach ridges of the southwest coast. On the mid and high altitude areas *Neonauclea-Sloanea* tall forest, *Garcinia-Elaeocarpus* mid-height forest, *Cyathea-Bambusa* scrub, and *Gulubia-Pandanus* (palm-pandan) vegetation are found.

An area of known timber potential exists at Tonolei Harbour in the southeast, and it is possible that stands of *Terminalia-Campnosperma* in swamp and riparian locations may have commercial value, although Heyligers observes that aerial photographs taken in 1962 revealed that *T. brassii* was dying in large areas of the Moila land system west of Boku and in swamps of the Luluai River in the southeast (Heyligers, CSIRO, 1967:137).

Although much forest remains on the plains and fans of Southwest Bougainville, much of it has been greatly disturbed by man's agricultural activities. It seems likely that its best future use will be to supply local needs rather than for commercial timber exploitation.

Land systems

The CSIRO study mapped Bougainville in terms of land systems (CSIRO, 1967). The categories present in the Buin subdistrict are shown in Fig.2.6 and the description of each is given in Appendix I.

Land use capability[1]

The CSIRO study assessed the land systems of the area in terms of land use capability. Capability classes are based on seven limiting factors, namely erosion, drainage, fertility, flooding, rockiness, stoniness and alkalinity. Table 2.2 shows the areas and percentages of land in each census division of the Southwest falling into each land use capability class. Fig.2.7 shows the areas falling within three capability classes in the eight census districts of Southwest Bougainville. The following land use capability classes are found.

(a) Lands suitable for cultivation with no limiting factors (Class I lands). There are no lands in this category in the area or in Bougainville, since the physiography is such that where the terrain is favourable the soils have limitations, and *vice versa* (Scott, CSIRO, 1967:171).

(b) Lands suitable for cultivation with minor limiting factors (Class II lands). Lands of high potential cover more than one-third of the area (1,184 km^2 with over 80 per cent in this category plus 127 km^2 with 20-50 per cent). They are suitable for the cultivation of crops adapted to the climatic conditions of the area. This category of land occurs mainly on plains of the Siwai and Buin land systems. The main limiting factor here is soil fertility, though slight drainage

---

[1] Material in this section is drawn from Scott (CSIRO, 1967: 168-73). I am indebted to Mr J. McAlpine, Division of Land Use Research, CSIRO, for providing the breakdown of land capability classes (Table 2.2) and land use (Table 2.3) by census division for the Buin subdistrict.

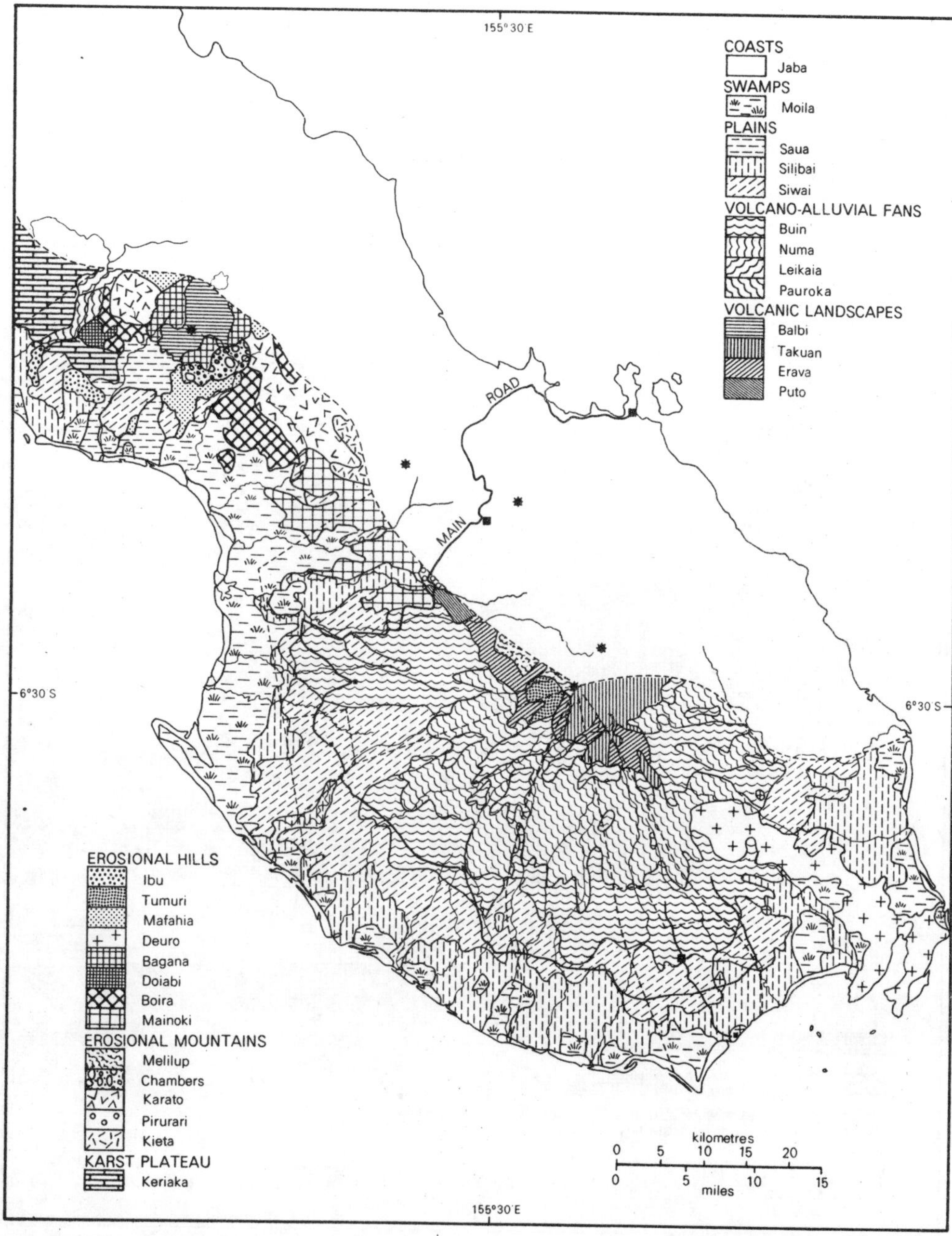

Figure 2.6 Land systems

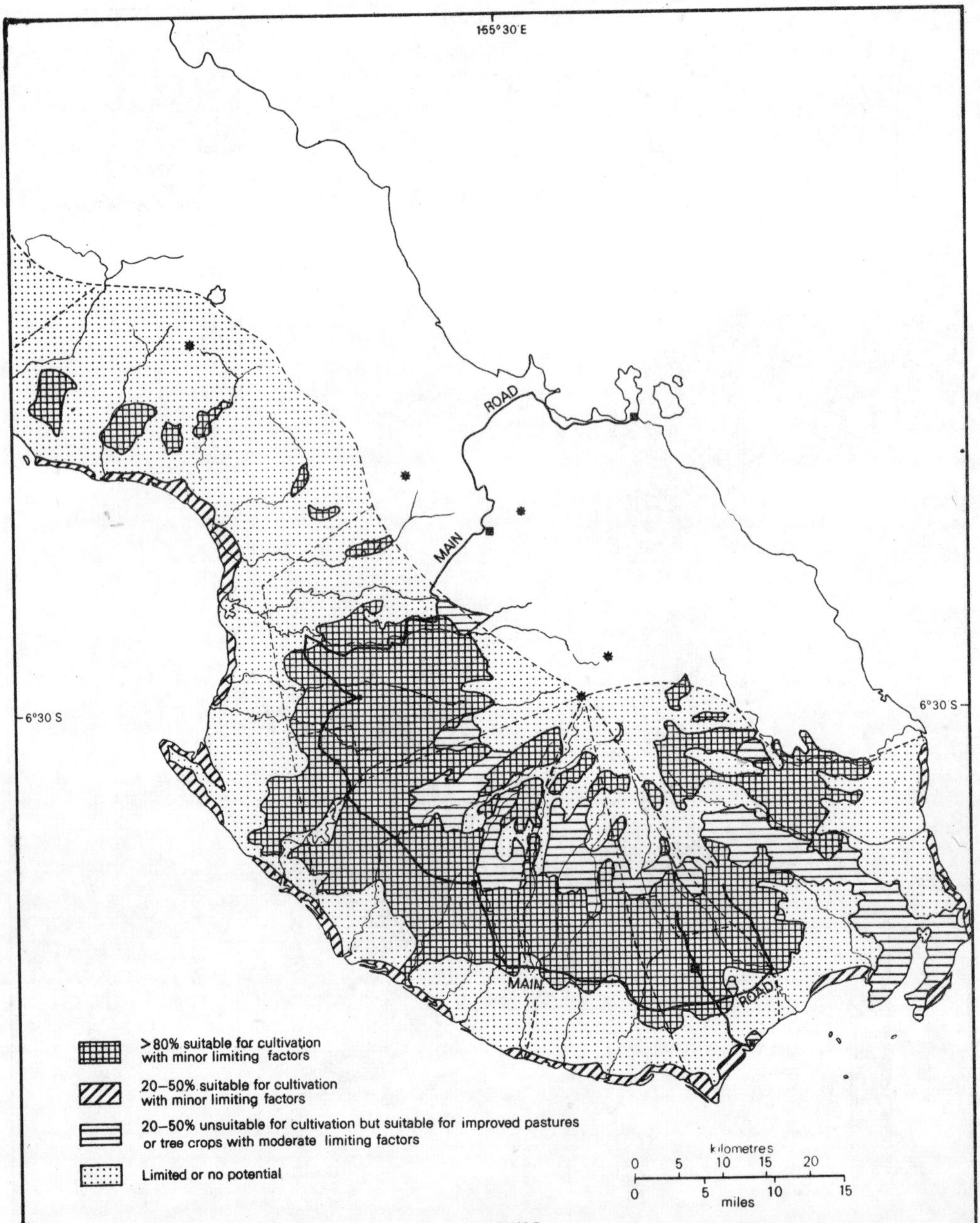

Figure 2.7 Land capability classes

Table 2.2

Areas of land in each capability class, by census division

	Class II		Class II		Class III		Class IV		Class IV		Class VI		Class VII		Class VIII		Class VIII		Total km^2	Comparison with total area in village directory
Proportion present	>80%		20-50%		20-50%		>80%		20-50%		20-50%		20-50%		>80%		50-80%			
Census division	km^2	%*	km^2	%*	km^2	%*	km^2	%*	km^2	%*	km^2	%*	km^2	%*	km^2	%*	km^2	%*		km^2
Banoni	78	7.3	73	6.8	9	0.8	76	7.0	63	5.9	nil	-	192	17.8	nil	-	585	54.4	1,086	1,018
Nagovisi	217	48.6	nil	-	nil	-	50	11.3	nil	-	18	4.0	40	9.0	48	10.7	73	16.4	446	479
Baitsi	78	94.5	nil	-	nil	-	5	5.5	nil	-	nil	-	nil	-	nil	-	nil	-	83	101
Siwai	340	53.1	11	1.8	nil	-	118	18.5	nil	-	63	9.8	2	0.3	63	9.8	43	6.7	640	827
Makis	101	32.9	9	3.0	nil	-	96	31.3	nil	-	55	18.1	5	1.5	23	7.4	18	5.8	307	257
Paubakei	103	53.8	11	5.9	nil	-	45	23.6	nil	-	6	3.5	nil	-	nil	-	25	13.1	190	141
Lugakei	58	43.0	nil	-	nil	-	nil	-	nil	-	18	13.1	11	8.4	48	35.5	nil	-	135	282
Kono	209	26.3	23	2.8	nil	-	164	20.6	nil	-	164	20.6	40	5.1	120	15.2	75	9.5	795	701
Total by class	1,184	32.2	127	3.5	9	0.2	554	15.1	63	1.7	324	8.8	290	7.9	302	8.2	819	22.3	3,674	3,806

* Percentage of the total land in each census division.

Note: Due to inaccuracies in the positioning of administrative boundaries on older maps there are discrepancies in these data. Areas of census divisions differ from those given in the Village Directory (TPNG, Department of District Administration, 1968) but the CSIRO figures are probably the more accurate.

Source: J. McAlpine, Division of Land Use Research, CSIRO, November 1972.

problems may occur on soils having ash-pan horizons in the Buin land system. Similar lands on the ridges and plateaux of the Buin land system may suffer from the additional hazard of erosion. Most areas have good access, though deep parallel valleys in the Buin land system are liable to restrict easy movement to the interfluves. For Class II lands occurring on the Jaba land system (where they occur on beach ridges) the main limiting factor is low soil fertility of the sandy soils. Access to them is made difficult by inundated swales, swamps on the inland margins, and surf beaches on the coast. A small area of Class II land occurs on Numa land system in the north of Banoni census division. There such lands are mostly confined to ridge crests where gentle slopes give rise to slight erosion hazards but where youthful lapillitic ash soils have low fertility. Access to these lands is difficult because they are encircled by hilly dissected terrain. There are also small areas of Class II lands in Pauroka, Takuan, Puto and Pirurari land systems, where erosion on steep slopes is the chief limiting factor, and to which access is difficult and easy movement restricted.

(c) Lands suitable for cultivation with moderate limiting factors (Class III lands). Only a very small area (9 km^2) on the Ibu land system in the northern Banoni census division falls in this category. Here moderate slopes lead to erosion hazards, and in addition the lapillitic ash soils are of low fertility. They are suitable for most types of agricultural production provided that intensive measures are taken against erosion. Access is generally easy, but easy movement tends to be restricted.

(d) Marginal lands for cultivation with major limiting factors (Class IV lands). Lands of moderate potential cover about one-sixth of the area of Buin sub-district (554 km^2, with over 80 per cent in this category on plains of the Silibai land system in the Southwest, and 63 km^2, with 20-50 per cent in this category on the Saua land system of northern Banoni census division). The plains of the Silibai land system are poorly drained and liable to be flooded for short periods at any time of the year. They are easily accessible in the Southwest. Movement within the Silibai land system is not easy, owing to the poor drainage conditions, flooding, water in drainage depressions, and the presence of many rivers. On the high plains of the Saua land system below Mt Bagana erosion risk is high and the extremely youthful ash soils are of low fertility. Access to and movement on these lands is restricted. Lands in this category have at present limited use because of the nature of the hazards.

(e) Lands suitable for improved pastures (Class V lands). No lands of this category occur in Southwest Bougainville.

(f) Lands suitable for improved pastures or tree crops, with moderate limiting factors (Class VI lands). Lands in this category cover 324 km^2 in the higher inland areas of Pauroka, Tumuri, Puto and Erava land systems in the Siwai and Makis census divisions and on the Deuro land system in Kono census division. They occur on slopes with moderate erosion hazards. Access is generally easy to moderate, except to those occurring on Erava land system.

(g) Lands suitable for tree crops, with severe limiting factors (Class VII lands). About one-thirteenth of Southwest Bougainville falls into this category (290 km^2, with 20-50 per cent of the land in this class), occurring on the Keriaka, Boira, Doiabi, Leikaia, and Mainoki land systems of northern Banoni census division, and on the high areas of Takuan land system in Makis, Lugakei and Kono census divisions. Generally such lands occur on slopes where there are severe erosion hazards. Access to them varies from easy to difficult.

(h) Lands unsuitable for any commercial land use (Class VIII lands). Some 1,121 km^2 of Southwest Bougainville falls into this category (844 km^2 with 50-80 per cent occurrence, and 302 km^2 with over 80 per cent occurrence). Most of this occurs on Moila land system (swamps), and Balbi, Bagana, Mafahia, Chambers, Karato (mountain) land systems in Banoni census division, and on Erava and Pauroka land systems in inland Nagovisi, Siwai, Makis, Lugakei and Kono census divisions. They occur mainly in swamps and river beds in the lowlands and on very steep slopes in upland areas. Varying amounts of such land are present in every land system.

Land use

The CSIRO study mapped and measured land use based on aerial photography taken in July 1962 at scales ranging from 1:60,000 to 1:73,000. The pattern for Southwest Bougainville is shown in Fig.2.8. Because of the small scale and indifferent quality of the photography an arbitrary grouping of land use categories had to be adopted. They are:

(1) Current subsistence agriculture. This comprises current (i.e. 1962) cultivation and clearing together with regrowth on gardens abandoned for less than about three years.

(2) Current subsistence agriculture and cash cropping. This comprises areas of current subsistence cultivation that were also known to include the major areas of cash cropping by the indigenous population.

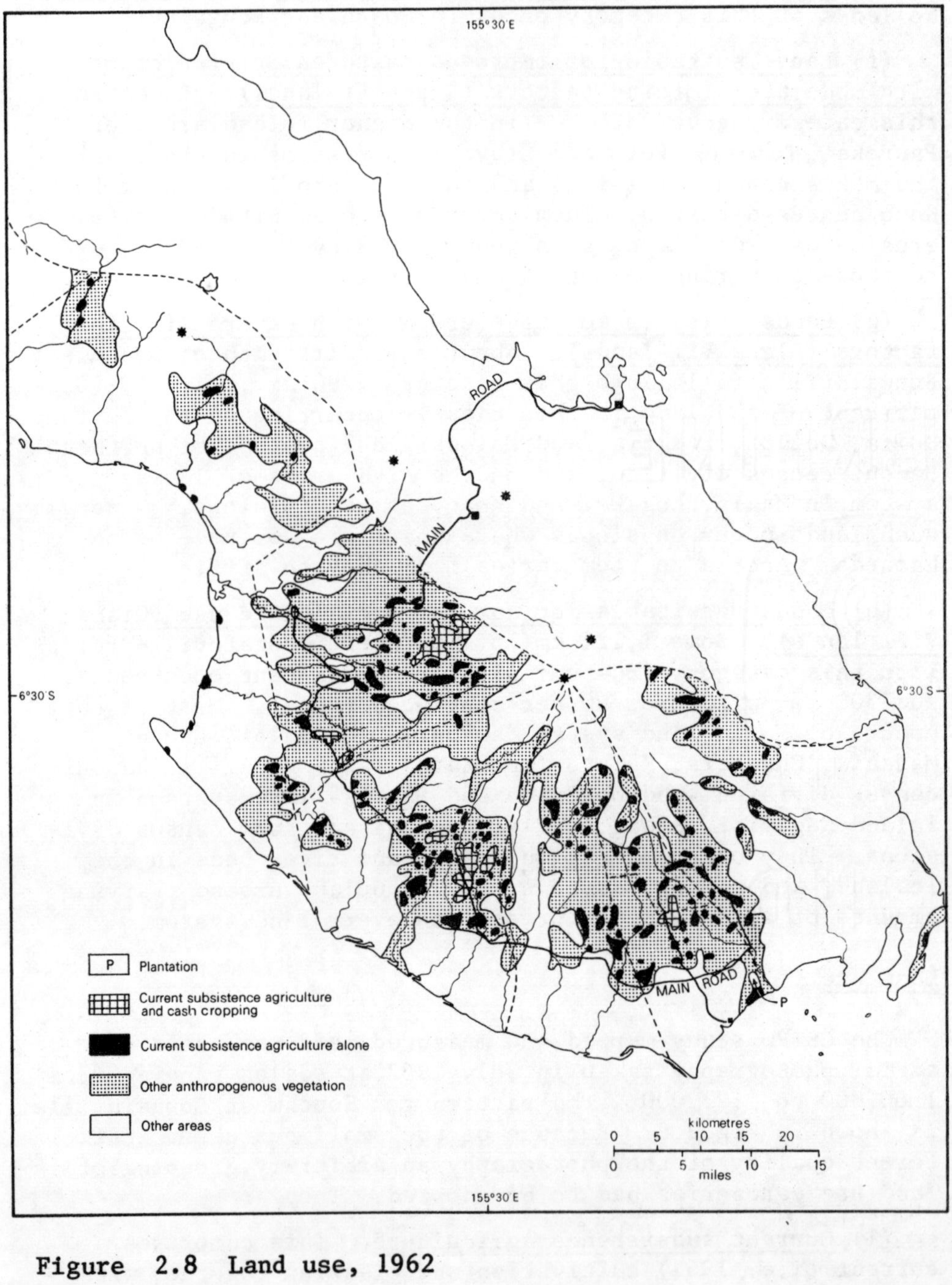

Figure 2.8 Land use, 1962

Table 2.3

Land use, 1962

Census division	Current subsistence agriculture alone		Current subsistence agriculture with cash cropping		Other anthropogenous vegetation		Other areas		Total	
	km^2	%	km^2	%	km^2	%	km^2	%	km^2	%
Banoni	2	0.2	nil	-	171	15.9	902	83.9	1,076	100.0
Nagovisi	5	1.0	11	2.5	292	65.4	139	31.0	447	99.9
Baitsi	2	2.8	2	2.8	58	72.3	18	22.0	80	99.9
Siwai	11	1.8	13	2.0	244	38.1	373	58.1	641	100.0
Makis	6	2.2	nil	-	71	23.0	229	74.8	307	100.0
Paubakei	5	2.4	nil	-	86	44.9	101	52.8	191	100.1
Lugakei	6	5.1	6	5.1	81	59.9	40	30.0	135	100.1
Kono	5	0.6	nil	-	192	24.0	602	75.4	798	100.0
Total	42	1.2	32	0.9	1,195	32.5	2,404	65.4	3,675	100.0

Note. See note to Table 2.2.

(3) Plantations. These are areas of non-indigenously owned plantations, of which only one occurs in Southwest Bougainville.

(4) Other anthropogenous vegetation. These areas were determined empirically by eliminating all areas currently showing no visible signs of current or previous use, that is all areas of primary or naturally occurring secondary vegetation. This mapping type indicates roughly the areas used for cultivation in the past 80-100 years. In populated areas it also indicates very broadly the maximum usable areas for subsistence cultivation.

Table 2.3 gives the areas in each category (about 1962) for the Buin subdistrict. The data show that in 1962 about one-third of the area had been in use for cultivation over the last approximately 100 years, while about two-thirds was apparently unused. Most of the sometime cultivated land was in the south-western census divisions, particularly in Buin and Siwai land systems. Banoni (84 per cent) and Kono (75 per cent) census divisions showed the greatest areas of unused land.

The areas shown as 'Other anthropogenous vegetation' (Fig. 2.8) are unlikely to have changed very much in Southwest Bougainville between 1962 and the present, though there may have been some increase in extent. Relatively small areas in each census division were in use for current subsistence agriculture. These areas will have changed in location since 1962, and may have increased in extent, although the rise of cash cropping is reputed to have caused a decline in the areas under garden cultivation, despite the increased population. McAlpine found an average of 0.4 acres per head per annum was under garden, based on a three-year garden rotation cycle (McAlpine, CSIRO, 1967:161).

Subsistence cultivation, today as in 1962, is based mainly on root crops. Before World War II taro (Colocasia esculenta) was the chief staple, but a fungus disease (Phytophthora colocasiae) infested the area in the late 1930s and 1940s, and virtually eliminated taro as a major subsistence crop. Sweet potato (Ipomoea batatas) and yams (Dioscorea sp.) are now the dominant subsistence crops, and are supplemented by bananas (Musa sapientum), sugar cane (Saccharum officinarum), pawpaw (Carica papaya), edible pit-pit (Setaria palmifolia), bread fruit (Artocarpus sp.), coconuts (Cocos nucifera), and cassava (Manihot sp.). Sago (Metroxylon salomonense) is collected and processed for food, and the plaited leaves used as thatch for houses. European vegetables and fruit such as maize, tomatoes, pineapples and beans are also grown, as is tobacco.

The land use mapping based on 1962 photography shows only very small areas in the category 'current subsistence agriculture and cash cropping'. The largest of these were 11 km^2 in Nagovisi census division, 13 km^2 in Siwai census division, and 6 km^2 in Lugakei census division. At about this time unsuccessful attempts had been made to introduce rice and peanuts as cash crops, and *robusta* coffee had been introduced, but cacao planting was just beginning. It was to prove a very successful introduction. It is likely that the areas indicated were of cacao, though coconuts were being grown as cash crops in Southwest Bougainville at the time. By 1972 there were 2.8 million cacao trees in Southwest Bougainville (see Chapter 3), and hence the area shown as under cash crops in 1962 had greatly increased by 1972.

Population

The enumerated village population of the eight census divisions of Southwest Bougainville was 31,070 in 1972. This represents 37.3 per cent of the total indigenous population of Bougainville District in 1972, and 34.9 per cent of the total population of the district in that year.

In addition there were relatively small non-indigenous and urban populations,[1] which would bring the Southwest Bougainville total to approximately 32,000 people in 1972. Details of these populations for 1969-70 are given in Table 2.4.

Table 2.4

Estimated non-indigenous and urban population, Buin subdistrict, 1969-70

	European	Asian	Mixed race	Indigenous	Total
Non-indigenous population	114	7	35	-	156
Urban (Buin town)	32	9	9	675	725
Total	146	16	44	675	881

Source: Bougainville District Annual Report, 1969-70, Appendix 12.

[1] The location of these populations includes Buin town; Boku Patrol Post; Konga; and the following missions: Torakina, Sovele, Moratona, Haisi, Tonu, Kataku, Piano, Tokaino, Turibairu, Kahili, Patapatuai, Tobago and Irunioku.

Table 2.5 shows the distribution of the 1972 population between the eight census divisions of Southwest Bougainville.

Table 2.5

Village population by census divisions, Buin subdistrict, 1972

Census division	Male	Female	Males per 100 females	Adult	Child-ren	Percent-age less than 15 yrs old	Total
Banoni	1,369	1,166	117	1,388	1,147	45.2	2,535
Nagovisi	3,630	3,195	114	3,154	3,671	53.8	6,825
Baitsi	458	392	117	440	410	48.2	850
Siwai	4,519	4,040	112	3,860	4,699	54.9	8,559
Makis	1,808	1,593	113	1,723	1,678	49.3	3,401
Lugakei	1,517	1,281	118	1,447	1,351	48.3	2,798
Paubakei	1,105	909	122	1,041	973	48.3	2,014
Kono	2,140	1,948	110	2,120	1,968	48.1	4,088
Total	16,546	14,524	114	15,173	15,897	51.2	31,070

Source: Village Population Registers for above census divisions, 1972.

The population is divided roughly equally between the northern, central and southern thirds of the area: Banoni, Nagovisi, and Baitsi census divisions having 10,210 persons; Siwai 8,559; and Makis, Lugakei, Paubakei and Kono census divisions 12,301. There are significantly more males than females in the population, the ratio being 114:100, and the range from 110:100 in Kono census division to 122:100 in Paubakei census division.[1] There is a high proportion of children in the population, 51.2 per cent being less than 15 years old, with Siwai and Nagovisi census divisions (54.9 and 53.8 per cent respectively) having the highest, Banoni census division (45.2 per cent) the lowest proportions. The relatively large number of children in the population has almost certainly been brought about by a reduction of infant mortality in the last decade, and implies a high potential for growth in the population.

[1] The ratios for Bougainville District were 121:100 in 1966 and 132:100 in 1971. The latter reflects the presence of men at Panguna-Arawa. The national ratios for the census years were 109:100 and 108:100 respectively (Bureau of Statistics, 20 December 1973).

Annual average growth rates for each census division between 1963[1] and 1972 are given in Table 2.6. These data produce an annual average rate of population increase for the last nine years of 4.09 per cent, with the highest values in Baitsi, Siwai and Banoni census divisions and the lowest in Kono and Paubakei. On the basis of this evidence there seems to be little doubt that the population is increasing at the rapid rate of about 4.0 per cent per annum in Southwest Bougainville.

Table 2.6

Population growth rates 1963-72 by census division

Census division	Population 1963	Population 1972	Annual average growth rate (%)
Banoni	1,490	2,215*	4.50
Nagovisi	4,888	6,825	3.78
Baitsi	544	850	5.08
Siwai	5,509	8,439**	4.85
Makis	2,453	3,401	3.70
Lugakei	1,975	2,798	3.95
Paubakei	1,475	2,014	3.52
Kono	3,021	4,088	3.42
Southwest Bougainville	21,355	30,630***	4.09

* Less two villages (Beteriopaia and Lesiopaia) not included in 1963.

** Less one village (Hino) not included in 1963.

*** Less the above three villages.

Source: Village Population Censuses, 1963 and 1972.

In Table 2.7 crude predictions[2] of the future indigenous population of Southwest Bougainville are given, using two rates, 3.0 and 4.0 per cent per annum. These projections indicate that unless more effective steps towards the control of population growth are taken than has hitherto been the

[1] 1963 data kindly supplied by Mr J. McAlpine, CSIRO.

[2] In making these projections no account has been taken of age structure or sex ratios of populations, or of other factors which could affect the results, since the village population data are not detailed or accurate enough to permit it.

Table 2.7

Projected indigenous populations by census divisions and Buin subdistrict, 1982 and 1992, at growth rates of 3.0 and 4.0 per cent per annum

Census division	1982		1992	
	at 3.0%	at 4.0%	at 3.0%	at 4.0%
Banoni	3,407	3,752	4,579	5,555
Nagovisi	9,173	10,103	12,328	14,954
Baitsi	1,142	1,258	1,535	1,862
Siwai	11,503	12,669	15,460	18,754
Makis	4,571	5,034	6,143	7,452
Lugakei	3,761	4,142	5,054	6,131
Paubakei	2,707	2,981	3,638	4,413
Kono	5,494	6,051	7,384	8,957
Total, Buin subdistrict	41,758	45,990	56,122	68,078

case Southwest Bougainville may have a population of between 56,000 and 68,000 by 1992. The implications of this on the economy and land use of the area are discussed in Chapter 3.

Village size. The population of Southwest Bougainville lives in nucleated villages, or at least are enumerated in such settlements for census purposes. A list of villages and their 1972 populations is given in Appendix II. In that year there were 230 villages in the eight census divisions of Southwest Bougainville. Table 2.8 shows the number of villages and their average size by census division.

The average village size for the whole area is 135 persons, ranging from 170 for the five large villages of Baitsi census division to 117 for the thirty-five villages of Kono census division. The absolute range is from 20 (Komai village, Kono census division) to 446 (Laguai village, Makis census division). Village sizes are not static since apart from changes due to natural increase or migration, larger villages occasionally split into two or more 'lines' with one group re-establishing at a new location. Small villages with an ageing and declining population may amalgamate with another village.

The modal size for the whole area is about 100-110 persons, with a secondary peak at about 190 persons. Two-thirds of all villages in the area have 70-160 persons, and a further 10 per cent have 180-200 persons.

Table 2.8

Number of villages and average village size, by census division, Southwest Bougainville, 1972 (including absentees)

Census division	Total population	Number of villages	Average village size
Banoni	2,535	21	121
Nagovisi	6,825	45	152
Baitsi	850	5	170
Siwai	8,559	65	132
Makis	3,401	23	148
Lugakei	2,798	21	133
Paubakei	2,014	15	134
Kono	4,088	35	117
Buin subdistrict	31,070	230	135

Source: Village censuses, 1972.

Distribution. The distribution of population by village size is given in Fig.2.9. The most noticeable feature of this distribution is the concentration of villages (87 per cent) on the plains and volcano-alluvial fans of the Siwai, Buin and Pauroka land systems (Table 2.9).

This concentration suggests that the type of environment offered by the Siwai, Buin and Pauroka land systems is most suited to the present economy of the population, and that the larger populations of the future will first seek space on these types of land. Fortunately, there are still considerable areas available within these land systems, but although some of this land is located in the accessible coastal lowlands, the remainder is on the higher mountain slopes of the interior, where access is at present difficult, slopes are steeper, the land more dissected, and rainfall higher.

Density. Table 2.10 gives crude population densities by census division for Southwest Bougainville. The overall population density in Buin subdistrict in 1972 was about 8 people per km^2, approximately the same as that for Bougainville Island in 1970. Within the subdistrict there were variations in density, Paubakei, Nagovisi and Makis census divisions having relatively high values, Lugakei, Siwai and Baitsi census divisions having moderate values, and Kono and Banoni census divisions having relatively low values.

The population density of Southwest Bougainville is lower than that of most inhabited parts of the Highlands, Maprik

Table 2.9

Distribution of villages by land system

Land system	Census division								Total	Per cent
	Banoni	Nagovisi	Baitsi	Siwai	Makis	Lugakei	Paubakei	Kono		
Coasts:										
Jaba	2			1	1			1	5	2.2
Swamps:										
Moila	4								4	1.7
Plains:										
Saua	1								1	0.4
Silibai	1	2		1	1		2	1	8	3.5
Siwai	5	4	5	24	3		4	5	50	21.7
Volcano-alluvial fans:										
Buin		34		25	13	7	9	17	105	45.7
Numa	1								1	0.4
Pauroka		4		14	5	14		8	45	19.6
Volcanic:										
Puto		1							1	0.4
Erosional hills:										
Ibu	1								1	0.4
Deuro								3	3	1.3
Boira	2								2	0.9
Erosional mountains:										
Karato	4								4	1.7
Total	21	45	5	65	23	21	15	35	230	99.9

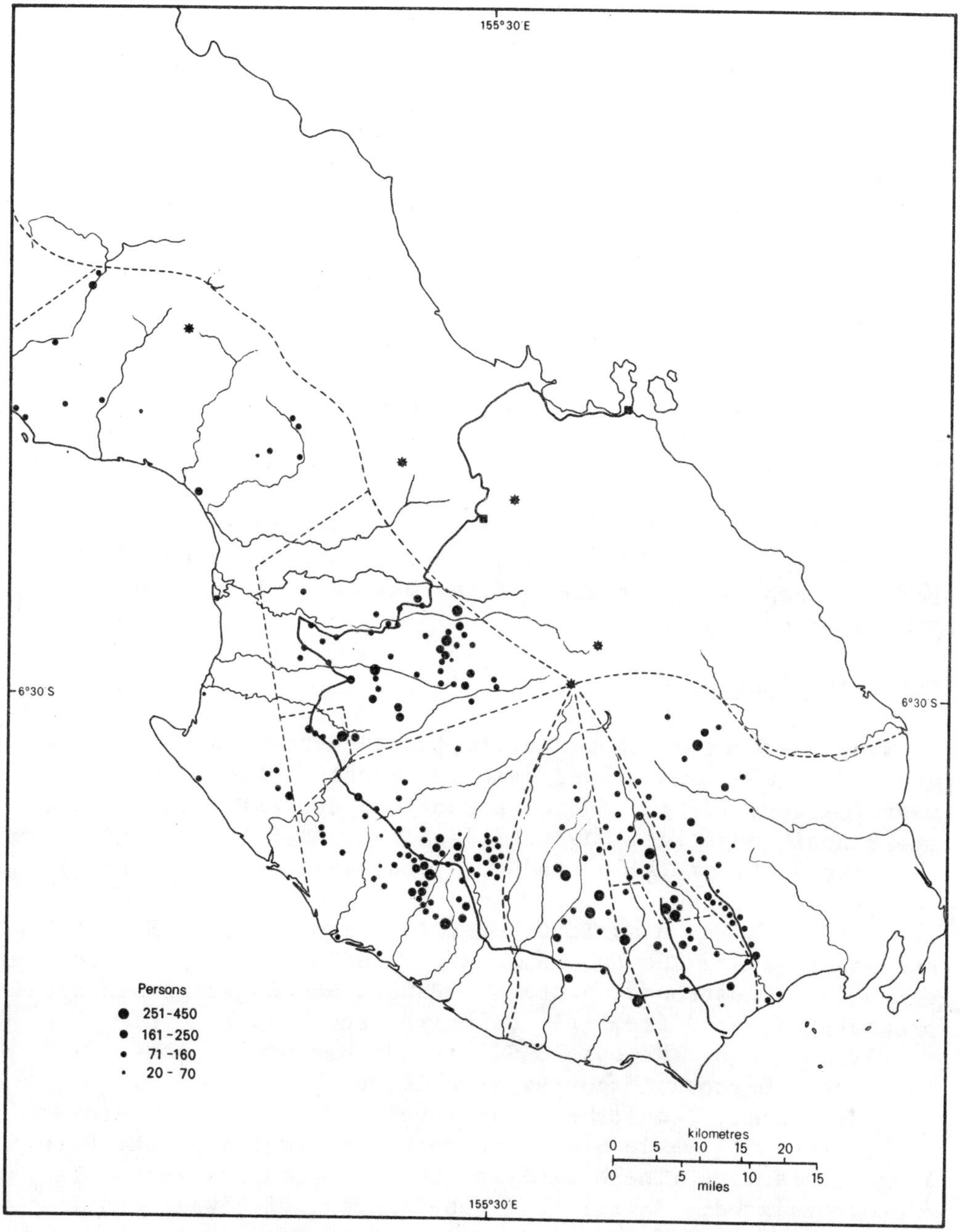

Figure 2.9 Village population, 1972

Table 2.10

Population densities by census division, 1972

Census division	Approximate area (km^2)	Total population	Density nos/km^2
Banoni	1,018	2,535	2.5
Nagovisi	479	6,825	14.2
Baitsi	101	850	8.4
Siwai	827	8,559	10.3
Makis	257	3,401	13.2
Lugakei	282	2,798	9.9
Paubakei	141	2,014	14.3
Kono	701	4,088	5.8
Buin subdistrict	3,806	31,070	8.2

area, and the Gazelle Peninsula, but higher than most of the rest of Papua New Guinea. The density concept is refined by relating population to land of various capability classes in Chapter 3.

Linguistic patterns

The languages of Bougainville Island were studied by Allen and Hurd (1963) and Oliver briefly reports those of the Southwest (Oliver, 1955). Those spoken in Southwest Bougainville are Banoni, Nagarige, Nagovis(i) (Allen and Hurd; called Sibbe by Oliver), Siwai (Allen and Hurd; called Motuna by Oliver), Buin (Allen and Hurd; called Rugara by Oliver, the speakers are sometimes known as Terei or Telei) and Baitsi. Banoni and Nagarige are the two members of the Banoni language family which is of Austronesian stock. Banoni and Nagarige speakers probably migrated from islands to the south at some time fairly long ago. Nagovisi, Siwai and Buin are of non-Austronesian stock; Nagovisi is a member of the Nasioi language family (the other member being Nasioi, spoken in the southeast), while Siwai and Buin are members of the Buin language family. The Nasioi and Buin language families are closely related. Baitsi is a sub-language of Siwai, and its speakers understand Nagovisi, due to their proximity to speakers of that language. Figure 2.10 indicates approximate language areas.

Melanesian Pidgin is widely understood and used throughout the area, and English is taught in primary schools and hence known to some degree by school children and by some adults.

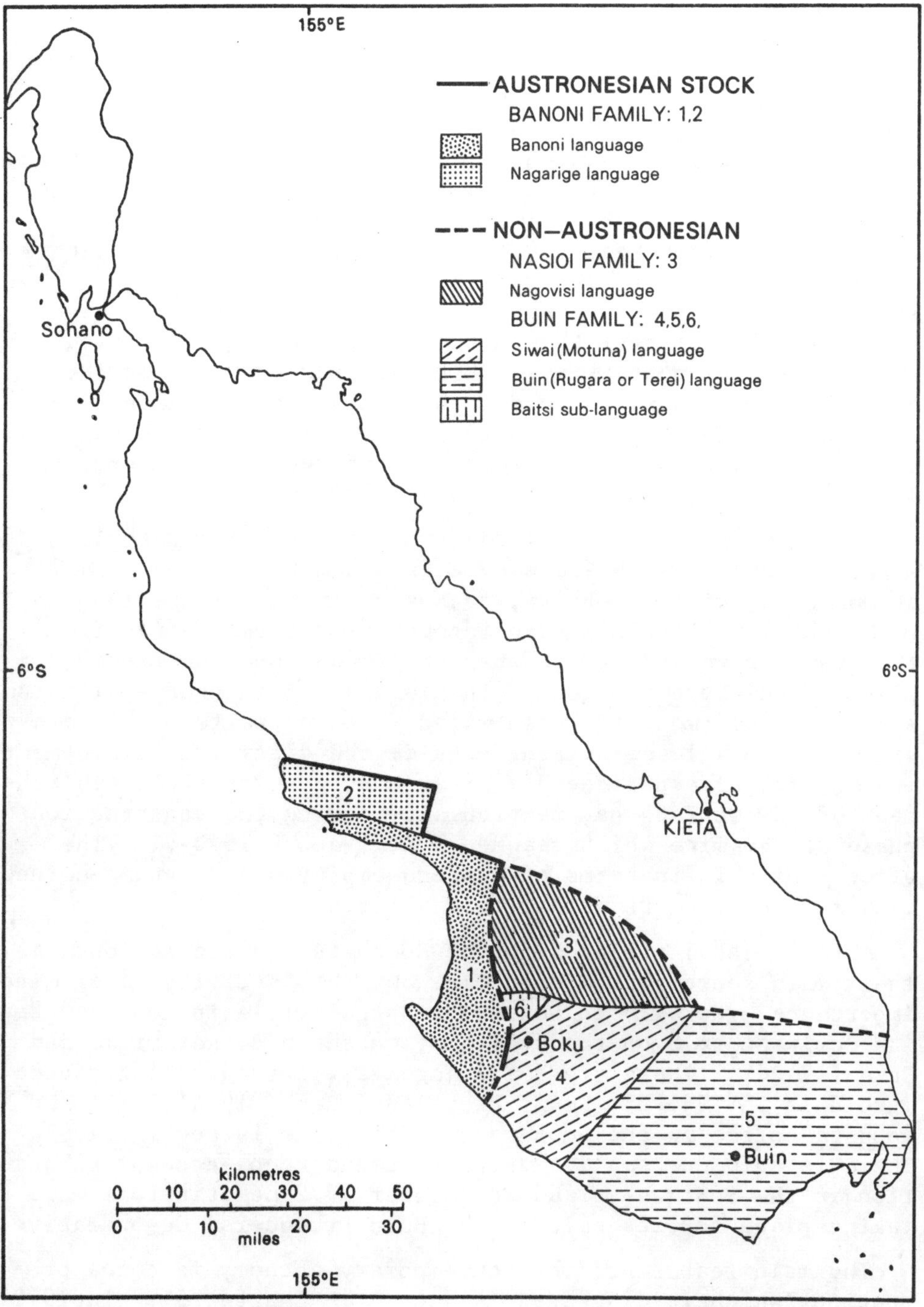

Figure 2.10 Languages

The economy

Until quite recently most of the population depended on village gardening (supplemented by pig-raising, fishing and hunting) for basic subsistence. The crops used are mentioned above, and the organisation of subsistence gardening has been described in general by McAlpine (CSIRO, 1967:160-4), and in detail for the Siwai by Oliver (1955:22-34) and for the Nagovisi by D. Mitchell (1971). Rice and peanuts were introduced (by the Administration) as cash crops in the late 1950s. Peanuts do not appear to have been very successful but are still grown as part of the subsistence crop complex. Rice enjoyed a brief popularity, 100 tons being produced in 1962-63 in Buin subdistrict and 80 tons in 1964-65 (Bougainville District Annual Report, 1964-65:14). Milling difficulties in that year contributed to a loss of confidence in the crop and little or none appears to have been grown since then.

Until the successful introduction of cacao (by Catholic missionaries about 1959 and by a returned plantation worker at Mosigetta village (Baitsi census division) before 1960 (D. Mitchell, 1971:218)), cash income was derived mainly from the manufacture and sale of baskets (made from the fibre of a plant of the *Lygodium* genus) in Siwai and Buin census divisions, and from employment for cash wages in other parts of Bougainville and to a lesser extent outside the district. Employment in wage labour experienced a great upsurge with the establishment of BCP at Panguna, particularly during the construction phase of the mine which reached a peak about 1970-71. The effects of this in terms of numbers employed and income earned is discussed in more detail in Chapter 4.

A few coastal villages in the Southwest depend on copra as their main source of cash income, but this activity is of minor importance generally in Southwest Bougainville in contrast to its dominant role on the coasts of northern Bougainville and Buka Island.[1] There are small acreages of rubber (introduced into Buin in 1964-65; Boug. Dist. Ann. Rep., 1964-65:15) and *robusta* coffee in the Southwest, but output is negligible. There is only one small expatriate-owned cocoa-coconut plantation in the Southwest, and in October 1972 negotiations were taking place for its sale to the Buin Producers' Co-operative.

The main feature of the contemporary economy is cocoa production (which is discussed in detail in Chapter 5). In 1972

---

[1] However, exceptionally high copra prices in 1973-74 may generate increased interest in copra in the Southwest.

there were an estimated 2,880,000 cacao trees in Southwest Bougainville, about 45 per cent in Banoni-Nagovisi-Baitsi, 17 per cent in Siwai and 37 per cent in Buin (DASF, Buin, typescript 23.8.72). Total production was 1,288 tons of cocoa in 1971-72 (ibid.). While this figure is only about one-fifth of the amount produced by plantations in other parts of Bougainville it represents about two-thirds of all indigenous-produced cocoa on the island. A proportion of the cocoa plantings are still immature, and new planting is continuing fairly strongly, so it is likely that cocoa production in the Southwest will continue to increase for some time. Increase in cocoa production appears to bear a direct relationship to the improvement of feeder roads (Smith, 1972:Fig.4, and see Chapter 3), hence investment in this form of development in the Southwest should receive high priority.

Environmental effects of the BCP mining development

BCP began exploration activities as Conzinc Riotinto of Australia (CRA) in central Bougainville in 1964, and made the decision to go ahead with the copper mining venture in 1967. The next four years saw a vast construction operation to bring the mine into production in 1972. As a single economic investment (some $A400 million), it is large by world standards and when considered in relation to the previous agriculturally based plantation/subsistence economy of a small island having a population of only some 70,000, it is enormous. Inevitably it has brought environmental, social, economic and political change. Perhaps the most important feature from the villagers' point of view has been, at least in the early stages, lack of knowledge about what was to happen and uncertainty as to the effects on them. The impact, however, has been felt more strongly on the eastern side of Bougainville Island, particularly in Kieta subdistrict.

In the west and Southwest, three main and one minor consequences of the BCP development may be identified, together with the more general political impacts which are shared by the whole island and to some extent with Papua New Guinea as a whole. The major consequences are the disposal of mine tailings in the Kawerong-Jaba River system, the new opportunities for wage labour (see Chapter 4), and the construction of the new road linking Southwest Bougainville with Kieta. The minor one is the creation of new vegetable marketing opportunities for a relatively small number of villagers in eastern Nagovisi (see Chapter 3).

Plate 2. The junction of the Pangala River (centre foreground) and the polluted Jaba River and the growing Jaba River delta in Empress Augusta Bay, Banoni census division, central west coast of Bougainville. The low parallel beach ridges of the Jaba land system are visible north of the river mouth.

The disposal of mine waste in the westward flowing Kawerong-Jaba river system (see Fig.2.11) has created a major environmental problem. The total input of sediment to the Kawerong-Jaba river system to March 1973 was 87,213,000 tonnes, of which about 68 per cent was carried into the sea, extending the delta of the Jaba River by about 1.3 km from the original coastline (Brown, 1973:6). The greatest build-up has occurred below the waste dump, in parts of the Kawerong river valley where eight metres of gravels have been laid down, and in the Jaba River below the Mokopolo sediment station where five metres of sand and silt have been deposited. About 300 hectares of lowlying land adjacent to the river have been affected by sedimentation, old river courses and ox-bow lakes have been filled in, and the spread of sediment has killed large areas of forest. The Jaba pump station, which pumps water from the upper Jaba River to the mine for industrial use, is seriously threatened by the excessive deposition and may have to be moved to prevent it being buried (Brown, 1972:736).

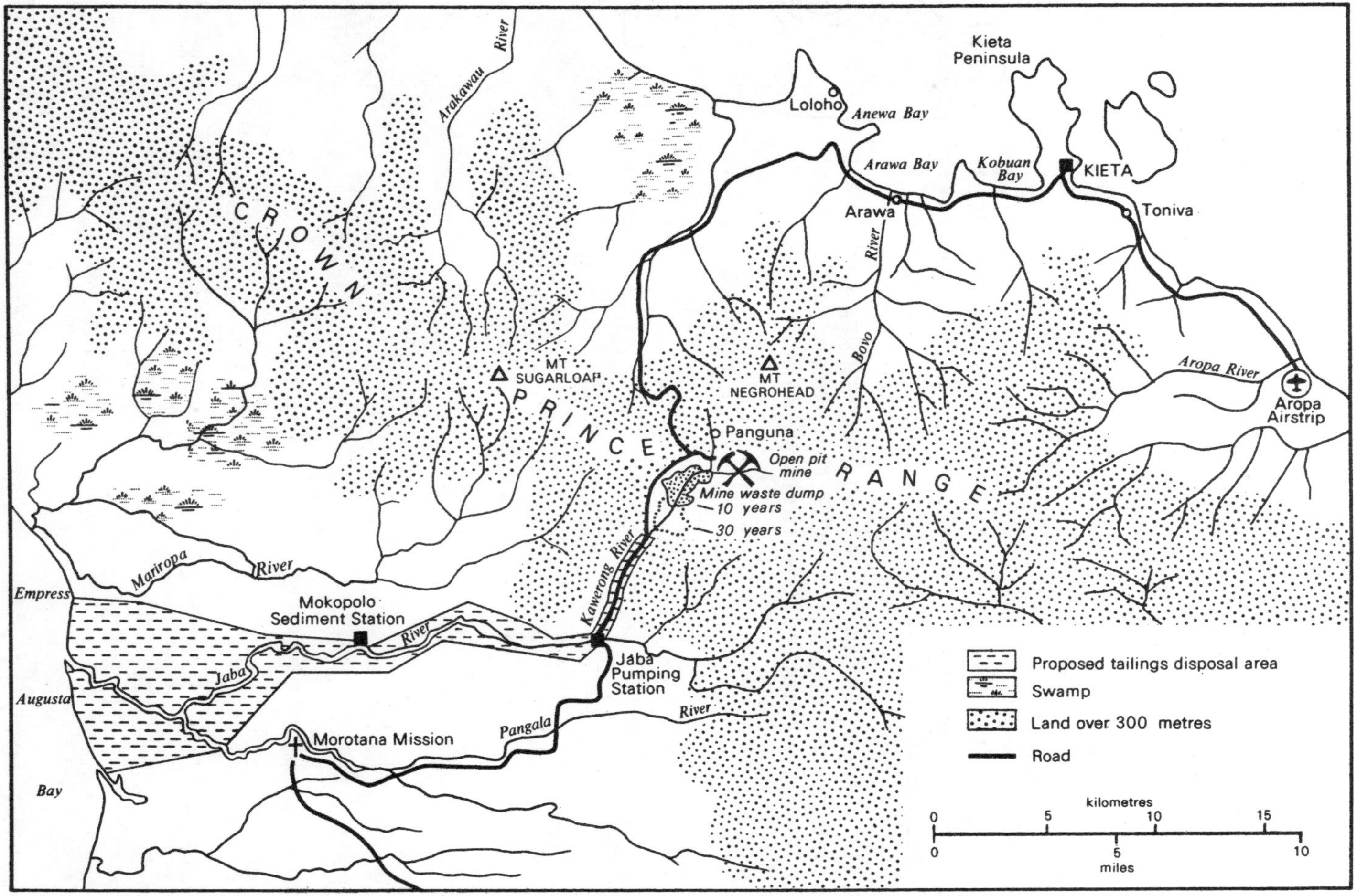

Figure 2.11 Tailings disposal area

This man-made deposition as well as chemical pollution has resulted in the destruction of the existing environment in the Kawerong and lower Jaba valleys, the killing of all fish in these rivers, and some destruction of garden or cash-cropped land. This has affected some villagers in Banoni and northern Nagovisi census divisions living immediately near the rivers, who formerly used the rivers for fishing, and whose land has been lost. BCP has paid compensation to the villagers for the damage they have sustained, and has assisted financially and with agricultural and business advice in the relocation of villages (including Jaba, at the Jaba River mouth), in the re-establishment of cocoa gardens, and in the establishment of at least one cattle project on the flats of the lower Jaba River. BCP also has plans to establish water buffalo on inland parts of the tailings affected area and, if successful, this will become another avenue of economic opportunity for local people. Despite these efforts, however, local people have naturally felt resentment at the destruction of their river and land resources, and at the time taken in deciding the amount of compensation they will receive.

Chapter 3

Economic activity

The main economic activity in Southwest Bougainville is agriculture. Subsistence agriculture still provides a basic stratum for virtually every family in the area, although home-produced food is now often supplemented by store-bought foodstuffs such as rice, canned meat, canned fish and other products purchased with cash earned from coca sales or from other sources. Because of the increased availability of cash, subsistence gardens are reported now to be smaller than formerly and to absorb less time and labour than used to be the case. Cocoa growing is the major feature of the indigenous economy of Southwest Bougainville, one of the two areas in Papua New Guinea where most of the indigenous produced cocoa of the country is grown. Copra and coffee are minor cash crops in the area. Basket-making is a minor craft industry which makes a significant contribution to income, particularly in Siwai and Buin, while the cash sale of fresh vegetables is important in the Kieta subdistrict and in small areas of Nagovisi and Kono census divisions. Indigenous cattle production is expected to become more important in future.

Cocoa

Cocoa was first introduced to the area about 1959 and was rapidly absorbed into the subsistence gardening economy. To establish a cocoa plot the land must be cleared, and a leguminous tree or coconuts planted as quick-growing shade. Cocoa seedlings are raised in individual containers in a shaded nursery, then planted about 4-5 m apart. This spacing will give about 620 trees per hectare of sole cocoa, or about 500 trees per hectare if interplanted with coconuts. Maintenance consists of occasional cutting of weed growth until the trees come into bearing after five years. The island is relatively free of cocoa pests (notably 'dieback' and *Pantorhytes*), but agricultural officers encourage growers to keep their cocoa plantings clear to discourage the spread of pests and diseases, and to enable increased production from each tree.

Harvesting consists of collecting the ripe pods and stripping the beans from them. The 'wet bean' must then be fermented for several days and eventually dried in a hot air drier, fired by wood or oil fuel to a controlled moisture content. These processes take place at fermentaries located either at the main co-operative purchasing centres or in small subsidiary fermentaries at village locations. The dry bean is then put into bags of about 64 kg weight and is ready for shipment. The co-operatives buy wet bean from the producers in small lots at a price which they determine according to the market prices they receive from the forwarding agents in Kieta. In 1972 Buin Society paid 0.5 cents/pound (1.1 cents/kg) and Siwai Society up to 0.6 cents/pound (1.3 cents/kg) for wet bean, when the market price was $490/ton ($498/tonne) for dry bean (File BD 12-4-1, Co-operatives Monthly Reports, Buin). About 98 per cent of all cocoa produced in Southwest Bougainville is sold through the co-operatives, the remainder being sold directly to traders (pers. comm. P. Wilson, DASF, Buin, 21.10.72).

Some measures of the past and projected growth of the indigenous cocoa industry are given in the tables that follow. The figures of plantings 1965-69 are illustrated in Fig.3.1. While they are likely to be only approximate they indicate a growth rate of 18 per cent per annum in total cocoa trees between 1965 and 1972. This represents a rapid degree of acceptance of an innovation. It seems that the returns from cocoa growing and the ease with which the crop could be incorporated into the existing agricultural system, together with favourable soil and climate conditions for the crop, generated a strong appeal to the people of Southwest Bougainville. Admittedly this expansion has taken place at a time when world prices (in terms of Australian currency) for cocoa were relatively favourable (Fig.3.2), despite several appreciations of the Australian dollar relative to sterling since late 1967.

It seems unlikely that new plantings will continue at such a high rate and Fig.3.1 indicates that planting rates have already apparently stabilised in the Siwai area (roughly 7 per cent increase per annum from 1965 to 1972). According to one source, the Buin Co-operative area appears to have experienced the greatest absolute growth in plantings (433,308) during 1965-69 (see Fig.3.1, inset) at a growth rate of 22 per cent per annum while the Bana Co-operative area had 260,512 new plantings at a growth rate of 23 per cent per annum. Data from another source available only for the Bana area confirm the rapid rate of expansion and indicate that it continued

Table 3.1

Number of cocoa trees planted, Southwest Bougainville, 1965-69, by co-operative society area

		Buin*	Siwai**	Bana***	Total
A. Mature 1965		261,743	319,699	307,241	888,683
New plantings	1965-66	41,773	35,286	35,811	112,870
	1966-67	50,578	19,119	25,456	95,153
	1967-68	154,839	28,610	79,301	262,750
	1968-69	186,118	32,935	119,944	567,753
Total	1969	695,051	435,649	567,753	1,698,453
Approx. total 1972		1,080,000	500,000	1,300,000	2,880,000
Annual percentage rate of increase	1965-72	22.4	6.6	22.9	18.3
B. Mature 1970-71		942,848	n.a.	n.a.	
New plantings	1971	98,403	n.a.	n.a.	
	1972	30,231	n.a.	n.a.	
	1973	70,085	n.a.	n.a.	
Total	1973	1,141,567	n.a.	n.a.	

* Census divisions: Makis, Lugakei, Paubakei, Kono.
** Census division: Siwai.
*** Census divisions: Nagovisi, Baitsi, Banoni.

Source: A. 1965-69: Department of Transport File 68/73(3) Bougainville District Roads General, Item 122 (Port Moresby). 1972: DASF, Buin (1972).
B. DASF, Buin (1973).

at least until 1970, as Table 3.2 and Fig 3.3 illustrate. On the basis of these data the Nagovisi census division experienced an even greater absolute amount of planting (910,085 new trees between 1965 and 1970), particularly in 1968 and 1969.

While they are only approximate, these figures indicate a growth rate averaging 43 per cent per annum for the Bana area between 1965 and 1970. Most of the expansion in the Bana area has occurred in the Nagovisi census division which therefore has at present the greatest potential for increase in production, provided adequate roading is available to allow removal of the crop.

Although the size of plantings seems high in absolute terms, it is not unrealistic when transformed to a trees and hectare

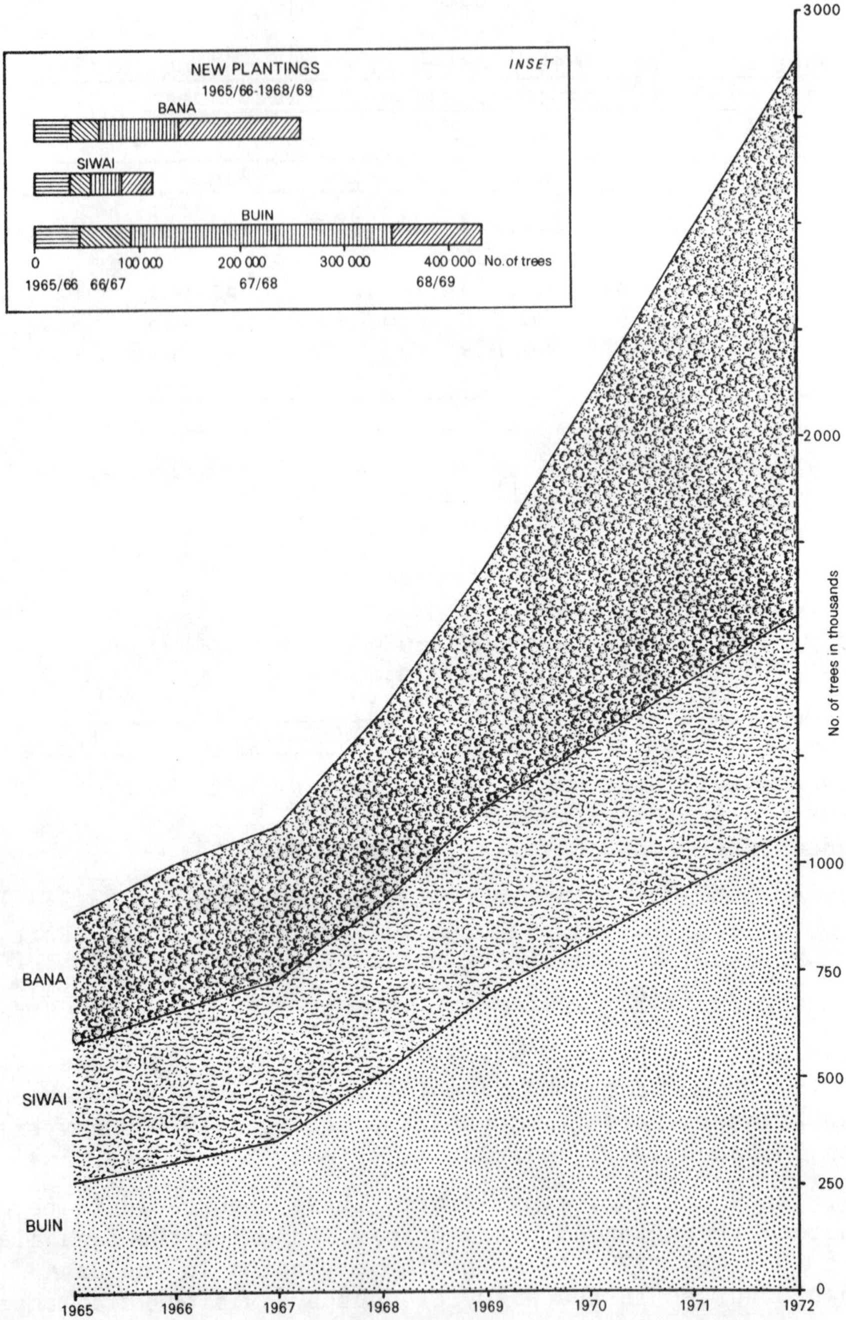

Figure 3.1 Cocoa plantings, 1965–72

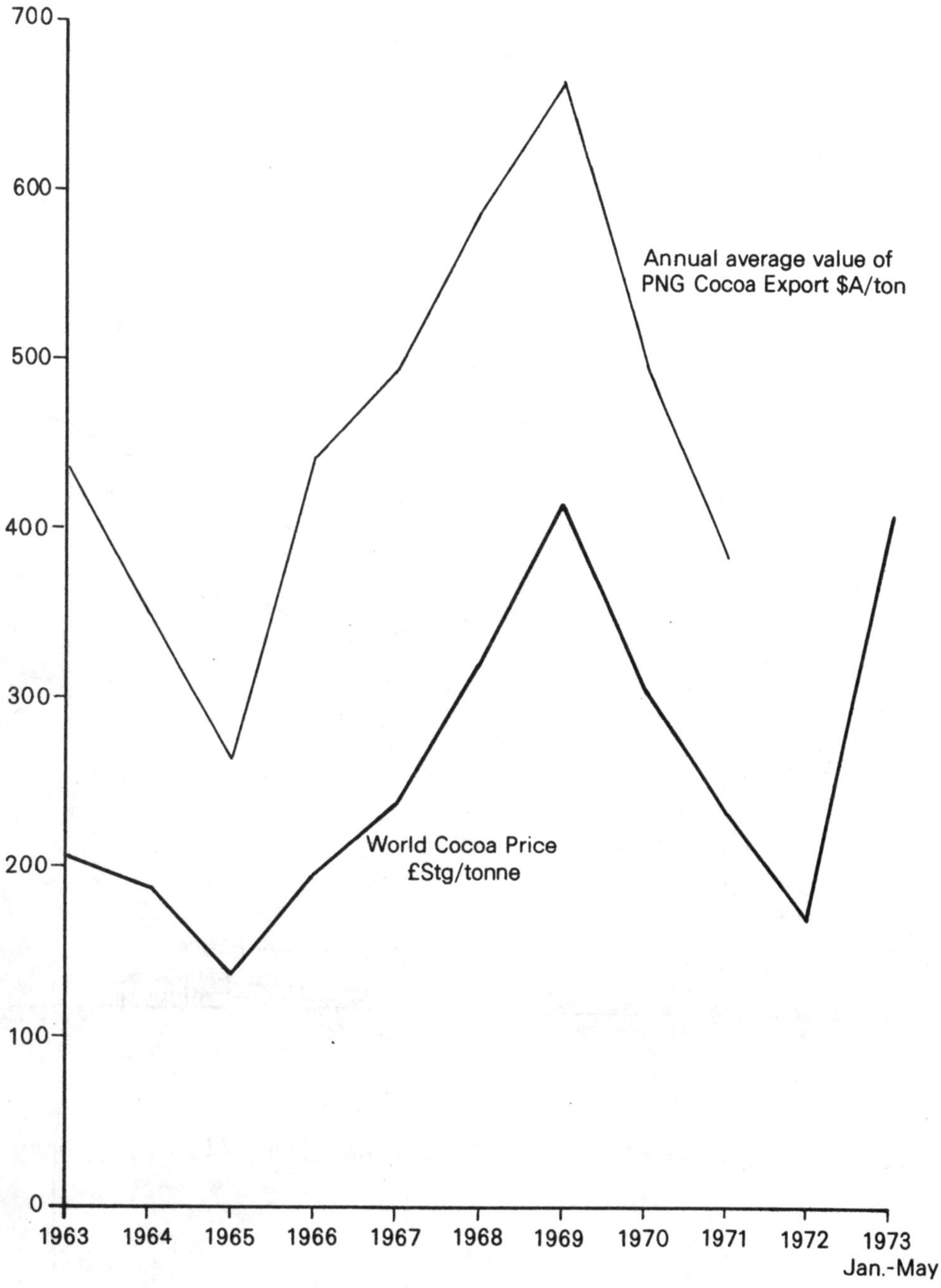

Figure 3.2 World cocoa price and annual average value of Papua New Guinea cocoa exports, 1963-73

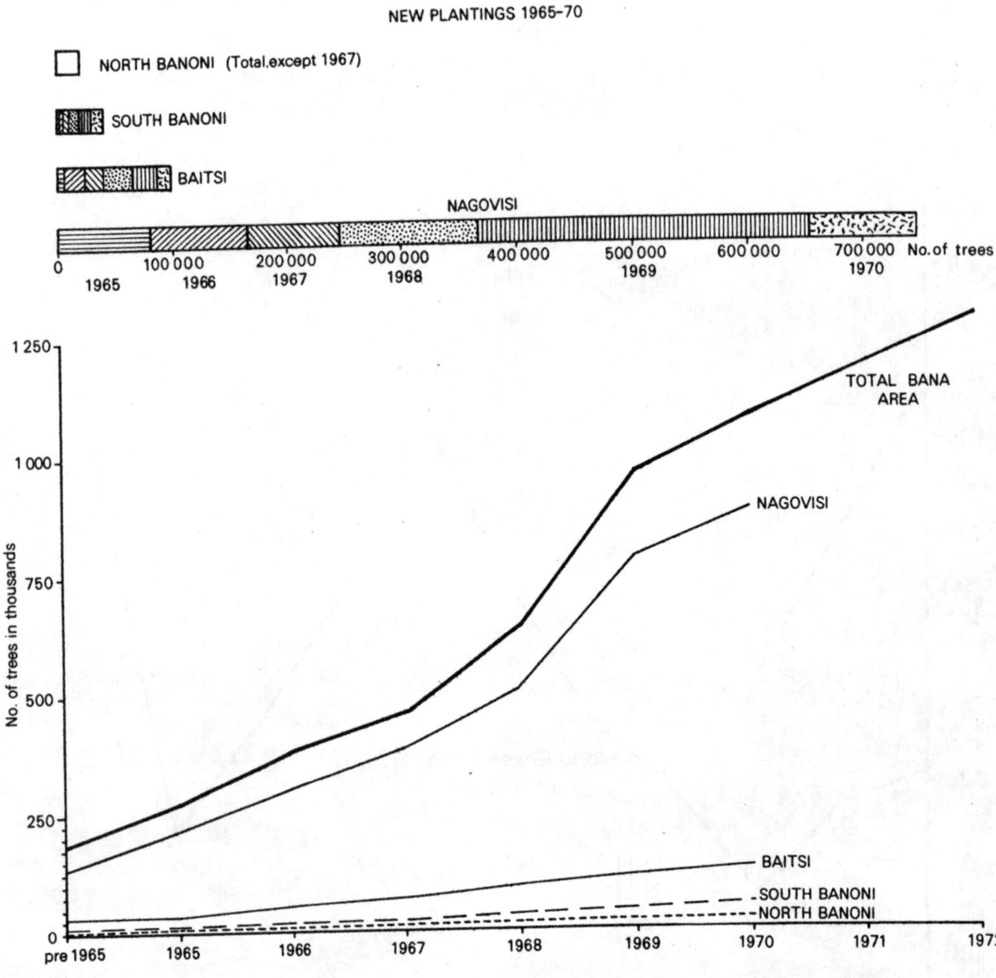

Figure 3.3 Cocoa plantings, Bana area, 1965-72

Table 3.2

Cocoa plantings, Bana area, 1965-70

		Nagovisi C.D.	Baitsi C.D.	South Banoni	North Banoni	Total Bana area
Mature		137,746	28,300	11,328	3,770	181,144
New plantings	1965	79,415	6,291	3,729	3,719	93,154
	1966	86,808	17,507	3,525	1,287	109,127
	1967	80,094	16,216	3,592	530	100,432
	1968	121,354	25,994	7,295	3,869	158,512
	1969	285,905	22,638	11,679	9,478	329,760
	1970	95,102	11,165	11,730	1,103	119,100
Total		886,484	128,111	52,878	23,756	1,091,229
Per cent increase of new plantings, 1965-70, of 1965 total		543.6	352.7	366.8	530.1	503.0
Annual percentage rate of increase		45.1	35.3	36.1	44.5	43.2

Source: Boku Patrol Report No.3 of 1970-71, Appendix C.

per capita basis, as Table 3.3 shows. While the per capita figures in Table 3.3 are only approximate, Siwai census division averages smaller plantings per capita than either Bana or Buin Society areas, and this is borne out in the village distribution map of cocoa plantings (Fig.3.4). All village plantings in Siwai are below 20,000 trees, whereas the range is extended to as high as 60,000 for one village in Nagovisi and 79,000 for one village in Makis census division.

Figure 3.4 also shows those areas in Southwest Bougainville which had dry weather road access only or which lacked road access entirely in 1972. Only the Siwai area and Paubakei census division had virtually all villages accessible by some sort of road, although many of these roads were of relatively poor quality. The inland areas of Makis, Lugakei, Kono, Nagovisi and all but the southern part of Banoni census divisions lacked road access or at best were served only by dry weather roads. Estimates of the number of cocoa trees planted in these poor access areas are given in Table 3.4.

A proportion of these trees are still immature so that the estimates of production given in Table 3.4 represent levels which could be reached when present plantings mature, i.e. by 1977. Future plantings would of course increase future potential production.

Table 3.3

Per capita cocoa plantings, Southwest Bougainville, c.1965 and c.1970

	Bana	Siwai	Buin	Southwest Bougainville
No. of trees 1965	307,241	319,699	261,743	888,683
Population (at date)	8,209 (1967 and 1970)	7,060 (1970)	9,766 (1965-6)	25,035
Trees/capita c.1965	37	45	27	36
Approx. ha/capita c.1965	0.37	0.44	0.27	0.35
No. of trees 1970	1,300,000	500,000	1,080,000	2,880,000
Population 1972	10,210	8,559	12,301	31,070
Trees/capita c.1970	127	58	88	93
Approx. ha/capita c.1970	1.26	0.57	0.86	0.91

Table 3.4

Estimated cocoa plantings and production in poor access areas, Southwest Bougainville, 1972

Area	No. of trees	Estimated potential annual production*(tonnes)
Buin		
Oria area	115,000	105
Tabago area	138,148	125
Inland Lugakei	61,787	56
Inland Makis	48,083	44
Bana		
Inland Nagovisi	c.500,000	453
South Banoni	40,000	37
Central Banoni Mtns	8,098	7
Banoni coast	2,840	3
Banoni Laruma River	10,175	9
Torokina Plains	2,298	2
Baitsi	20,000	18
Total	946,429	859

* Estimated on basis of 0.91 kg/tree dry bean yield.

Source: No. of trees: Information provided by DASF, Buin 23.8.72, and Boku Patrol Report No.3 of 1970-71, App.B. Production: calculated.

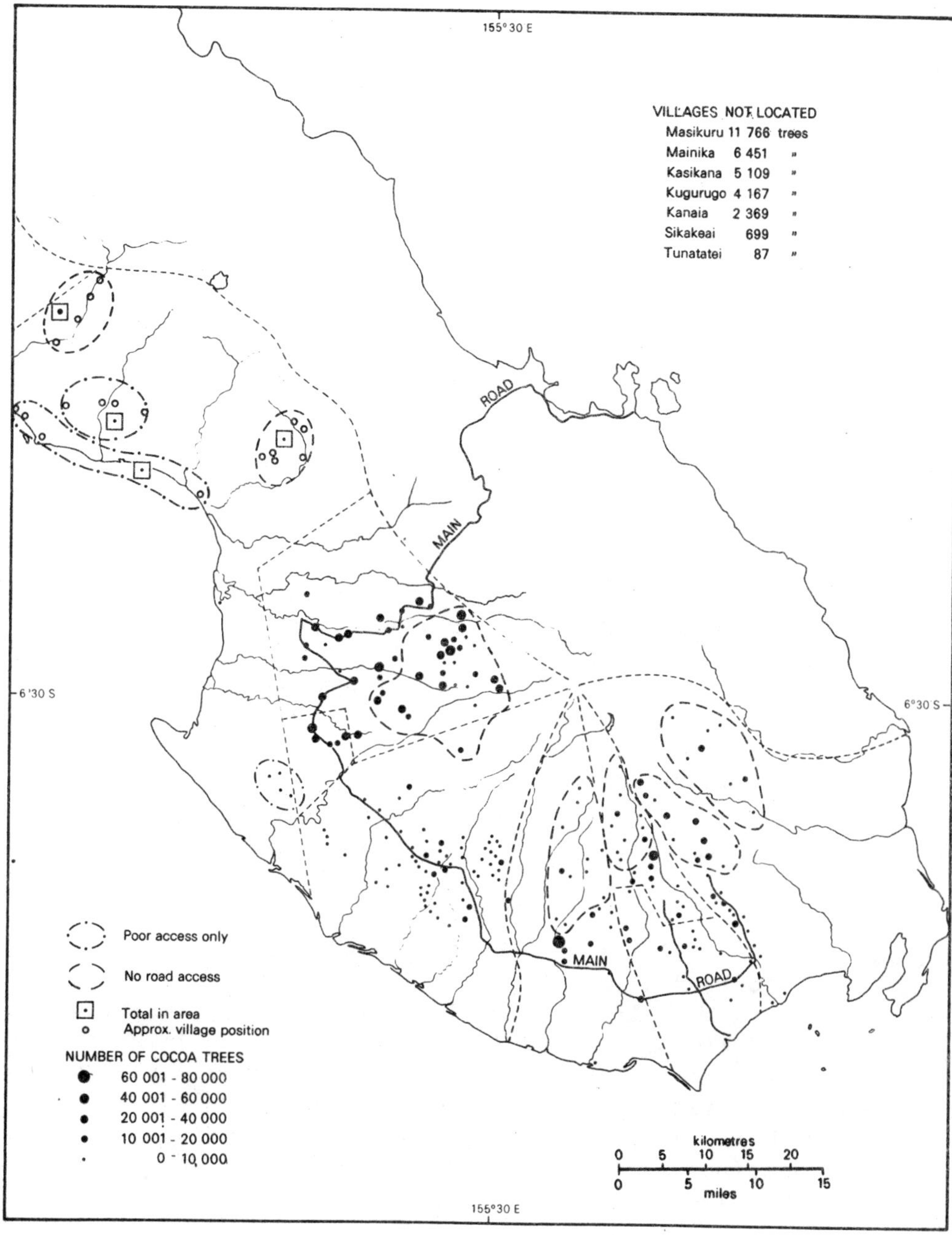

Figure 3.4 Cocoa plantings by village, 1970

There is evidence to suggest that improved road access will increase production. The upgrading of the section of road from Moratona to Sikoreva in Nagovisi census division in 1971 appears to be the main reason why the tonnage of cocoa produced by the Bana Co-operative Society increased by 28 per cent from 227 tonnes in 1970-71 to 291 tonnes in 1971-72. The production from the mature trees in 1971-72 of the nine villages served by this road would have accounted for about 42 tonnes of the 64 tonnes increase.

Available data on past and projected production of cocoa (dry bean) are given in Table 3.5 and illustrated in Fig.3.5. This table and graph show a steady increase in the actual amounts of cocoa produced by indigenous growers both in Southwest Bougainville from 1966-67 to 1971-72 and in Bougainville Island as a whole from 1963-64 to 1970-71. Southwest Bougainville's contribution has increased from about one-fifth to nearly one-half of indigenous-produced cocoa between 1966-67 and 1970-71. It seems likely that this share will be at least maintained and could be slightly increased by 1975-76.

Figure 3.5 also shows that the contribution of indigenous-produced cocoa _vis-à-vis_ plantation-produced cocoa is steadily increasing for Bougainville as a whole. Plantation cocoa increased rapidly from about 1960 to 1965 and again from 1966 to 1968, but appears to be levelling off since that date. Production of indigenous cocoa started later and shows its greatest rate of increase after 1968.

Within Southwest Bougainville there have been some differences between actual and projected production between 1969-70 and 1971-72.[1] In the Siwai and Bana areas the actual production in 1970-71 was less than that projected, possibly because of cyclones; in 1971-72 only Bana production was lower than the projection. But in Buin, actual production exceeded the projections in both 1970-71 and 1971-72. The figures used in Table 3.5 and Fig.3.5 for projected production in 1975-76 were estimated on the basis of production from 1972 plantings at a dry bean yield of 0.91kg/tree, the figure currently achieved in Buin from indigenous plantings.[2] This also assumes that present plantings will have reasonable access by that date, so that the full potential can be achieved from the areas

---

[1] The projections for 1969-70 to 1974-75 must have been made about 1970.

[2] A similar yield is achieved by indigenous smallholder cocoa producers in the Gazelle Peninsula.

Table 3.5

<u>Cocoa production, Southwest Bougainville, 1967-68 to 1974-75</u>
(tonnes)

	Buin		Siwai		Bana		Other		Total	
	Actual	Projected	Actual	Projected	Actual	Projected	Actual	Projected	Actual	Projected
1966-67									151	
1967-68	141	-	170	-	169	-			480	
1968-69	179	-	224	-	254	-			656	
1969-70	165	249		277		341				868
1970-71	373	322	258	324	227	400			858	1,046
	(356)		(244)		(254)				(853)	
1971-72	597	379	390	362	291	448	30(Oria)		1,309	1,189
	(597)		(314)		(324)				(1,235)	
1972-73		484		392		491				1,367
1973-74		568		419		534				1,623
1974-75		674		440		554				1,667
1975-76		979*		453*		1,180*				2,612*

* Estimated from yield of 0.91 kg dry bean/tree from total of 2,880,000 trees.

<u>Source</u>: Actual: DASF, Buin, and Smith, 1972:11. Alternative figures (in brackets) for 1970-71 and 1971-72 are from Co-operatives Development Officer, Buin, Annual Report, Buin area, 1972 (originally in tons).
Projected: Department of Transport File 68/73(3), Item 122, Port Moresby (originally in tons).

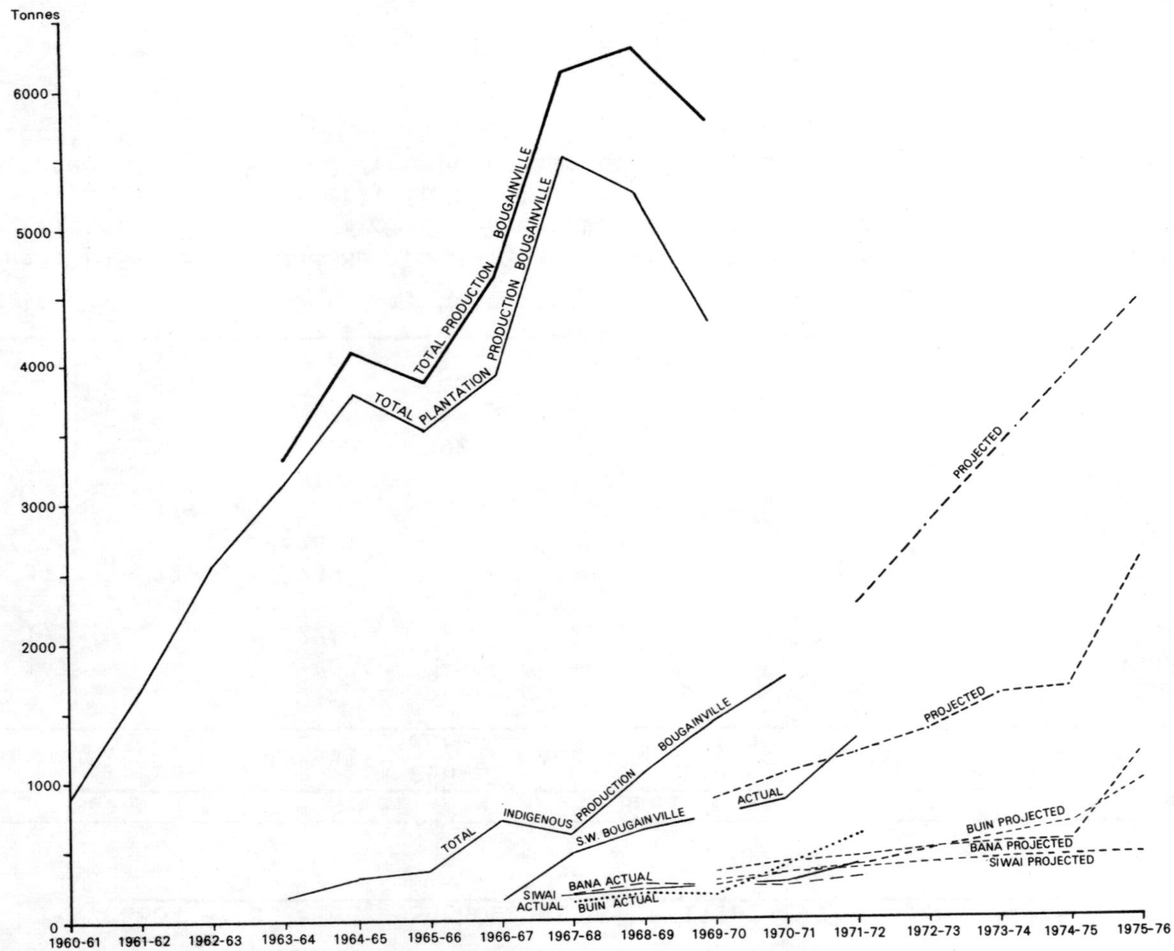

Figure 3.5 Volume of cocoa production, 1960/61-1975/76

shown in Fig.3.4 as having inadequate access. This accounts for the sudden upward turn between 1974-75 and 1975-76 in the projected figures for Buin, Bana and Southwest Bougainville. This is unlikely to represent reality in the event, a gradual continuing increase over the three preceding years being more likely, as feeder roads are gradually built into the areas of poor access.

The present position seems to be that the Siwai area has already stabilised at a lower planting rate and smaller per capita holdings than the other two areas, and hence its production will continue to increase relatively slowly. The Buin area has experienced a surge of planting which is now beginning to level out, hence its production may continue to rise fairly rapidly before also levelling out in the late seventies. In the Bana area, particularly Nagovisi census division, planting is continuing at a rapid rate, and the effects of this will be seen in a more rapid increase in production in the mid- and late seventies. All of these predictions assume continued improvement in quality and extent of feeder roads, and this, combined with pressure from a steadily growing population to plant new land as increasing numbers of young people enter the workforce, may continue to generate new plantings. The opinion has often been expressed that insufficiency of labour will slow down both planting and production in the next few years, and this may well occur. This question will be further examined in Chapter 4.

Virtually all marketing of cocoa in Southwest Bougainville is done by the three producers' co-operative societies, namely Buin Producers' Co-operative Ltd, Siwai Co-operative Producers' Society Ltd, and Bana Producers' Co-operative Society Ltd. The three societies buy wet bean from the producers and after drying it at their central fermentaries at Beretemba, Konga and Buin, or at subsidiary fermentaries, sell it as dry bean to forwarding agents at Kieta. A marketing co-operative was formed in 1972 with share capital subscribed by the producer co-operatives of south Bougainville, to sell cocoa direct to overseas markets.

As an illustration of seasonal fluctuation of wet bean production the figures for Siwai Society's monthly purchases for 1971 and part of 1972 are given in Table 3.6 and illustrated in Fig.3.6. The pattern of purchases by Bana Society for the same periods is similar.

There are usually two peak periods or flushes during the year, about May-June and September. The subsidiary fermentaries may only be used during flush periods.

Plate 3. Weighing bags of wet cocoa beans brought in for sale by Nagovisi villagers at Bana Producers' Co-operative, Beretemba

Table 3.6

Wet bean purchases, Siwai Society, 1971 and January-August 1972

	1971	1972
	tonnes	tonnes
January	25.032	49.955
February	26.125	54.092
March	45.950	80.901
April	59.306	96.092
May	122.274	91.911
June	85.744	79.867
July	43.682	31.313*
August	64.732	56.590
September	72.468	
October	38.769	
November	28.992	
December	36.990	

* The low figure is partly a consequence of the cyclone of May 1972 which depressed production.

Source: Siwai Co-operative Producers' Society (originally in pounds).

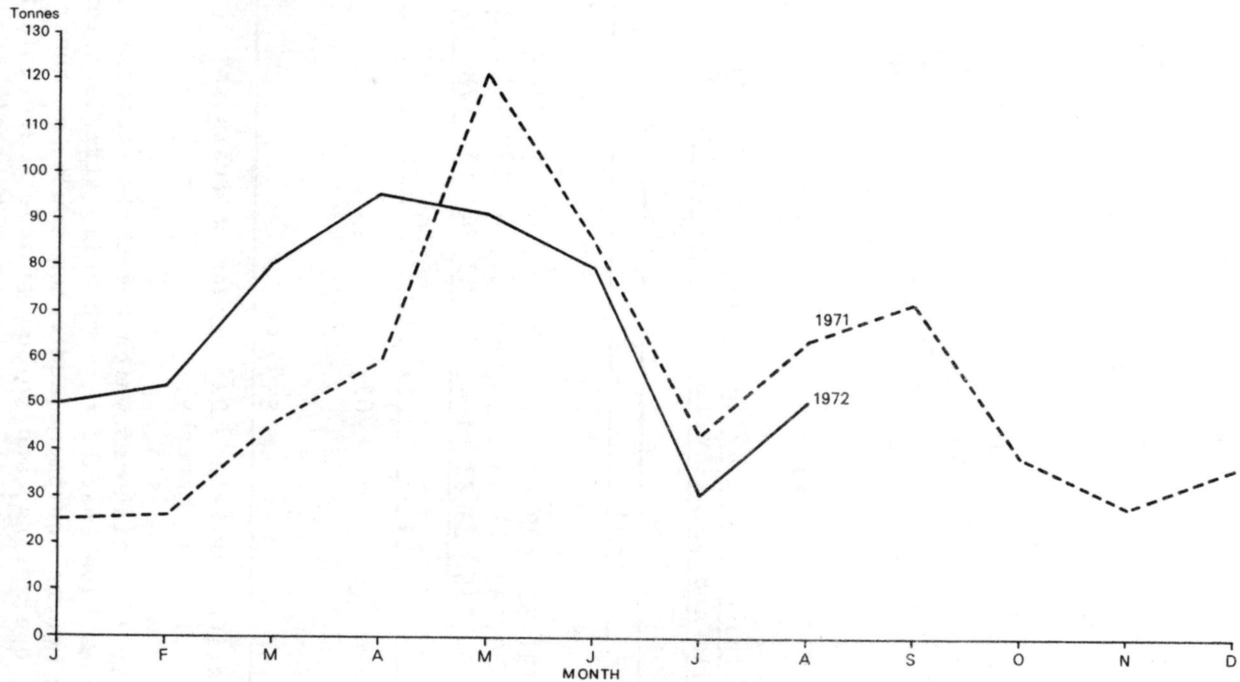

Figure 3.6 Wet bean purchases, Siwai Co-operative Producers' Society 1971 and part of 1972

The cocoa sales figures for three years, 1968-69 to 1970-71, for the three societies and Bougainville are given in Table 3.7 and Fig. 3.7. Nearly three-quarters of the value of co-operative societies' sales of cocoa (which accounts for a large proportion of all indigenous produced cocoa in Papua New Guinea) is provided by Bougainville Island, and more than half of this comes from Southwest Bougainville. The Southwest Bougainville total was roughly equally divided between the three societies in the years 1968-69 to 1970-71, but on the basis of new plantings and anticipated production discussed above it can be expected that the Bana Society's share will in future exceed that of Siwai. The larger share of Siwai in 1968-69 may reflect the presence of a network of passable feeder roads in that area compared with their relative absence in the other two areas, but could result from a variety of other causes, including variable weather conditions and earlier maturity of plantings as shown in Table 3.3.

Table 3.7

Value of sales of cocoa by co-operatives, Southwest Bougainville and Bougainville, 1968-69 to 1970-71

Society	1968-69		1969-70		1970-71	
	$	%	$	%	$	%
Bana	91,354	28.87	114,713	29.10	116,690	28.65
Siwai	130,787	41.33	114,960	29.16	137,762	33.82
Buin	94,336	29.81	164,523	41.74	152,878	37.53
S.W. Bougainville	316,477	100.01	394,196	100.00	407,330	100.00
Bougainville	514,760		707,253		n.a.	
S.W. Bougainville as per cent of total	58.4		55.7		n.a.	

Source: Co-operatives Division, Department of Business Development, Port Moresby.

The relatively large incomes which the three societies have enjoyed from cocoa have enabled them to build efficient fermentaries using modern equipment and cocoa stores at their main centres of operations, and to buy vehicles, and this in turn has helped them provide better services to their producer members. It has also enabled them to consider diversification into other commercial activities.

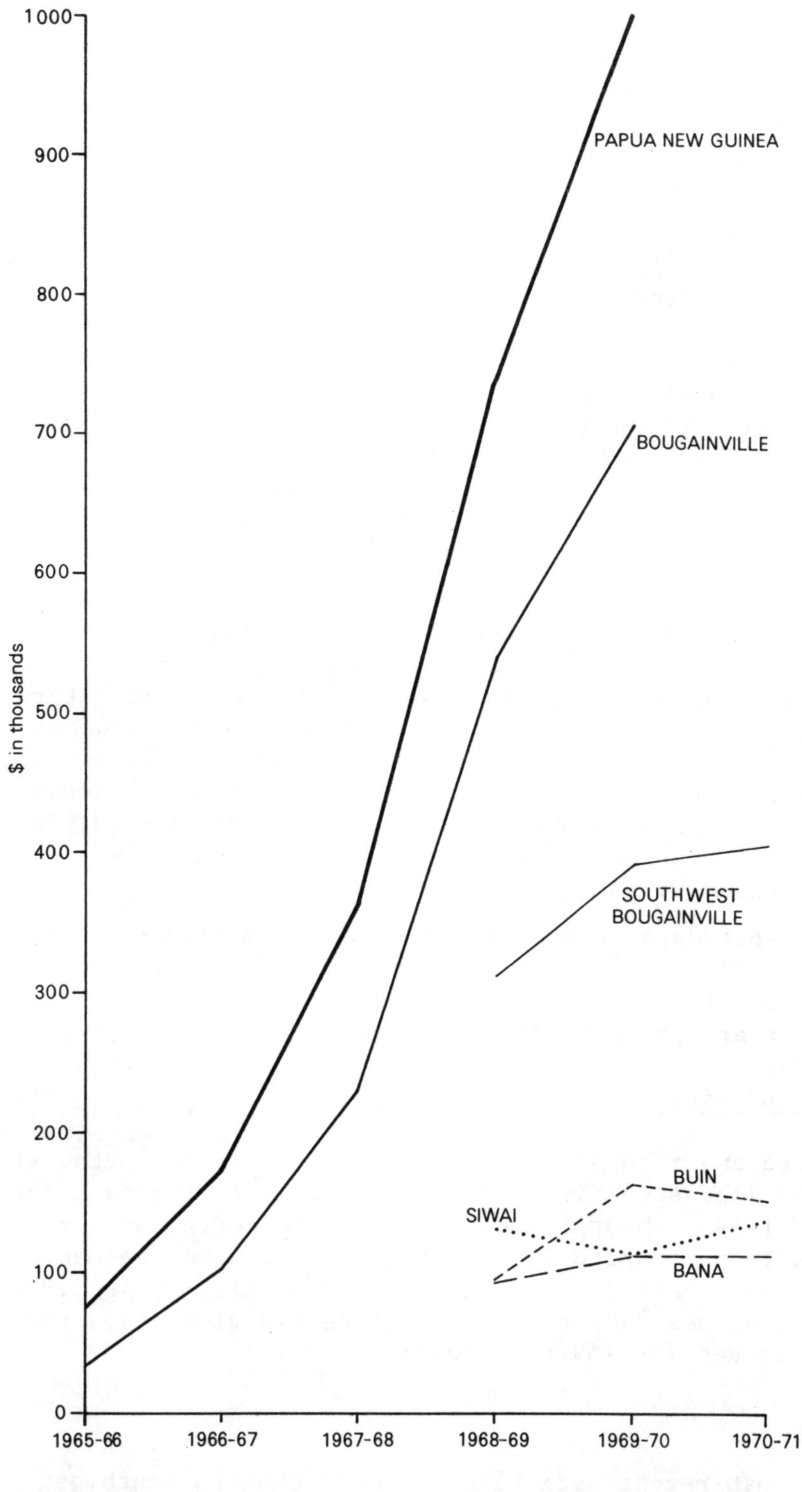

Figure 3.7 Value of cocoa sales (cooperatives)

Copra

Copra production is much less important than cocoa in Southwest Bougainville, and much less is produced by indigenous growers there than in the north and east of Bougainville Island. The physical reasons thought to contribute to the lower yields achieved in Southwest Bougainville have been suggested in Chapter 2.

There were some 649,964 coconut palms in Southwest Bougainville about 1970, of which 373,165 were mature. Production about 1970 was reported to be 119 tonnes which is negligible compared with about 20,320 tonnes per annum for the whole island. Production could double by about 1975-76 if present rates of increase of production continue, and if the cause of the low yields can be resolved, production could reach 500-1,000 tonnes per annum (Dept of Transport File 68/73(3), Item 122). The high copra price of 1973 could well induce increased production.

Plantings appear to be concentrated in Siwai and Buin Society areas. A few coastal villages are entirely dependent on copra as a cash crop, but most of the palms are interplanted with cocoa on Buin and Siwai land systems of the lowlands. Siwai census division was reported to have 202,500 palms in 1970 (Konga Patrol Report No.15 of 1970-71), which would give the four southern census divisions about 350,000-400,000 palms, with the remainder in Nagovisi and coastal villages of Banoni census division.

Available data on volume and value of copra production are summarised in Table 3.8. This refers only to co-operatives' production and sales, and it is possible that perhaps another 60 tonnes are produced and sold privately.

Coffee and rice

Coffee and rice are very minor cash crops in Southwest Bougainville, significant at present only as introductions which failed. About 5 tonnes of *robusta* coffee were reported to have been produced in 1970 in the Bana area, valued at \$2,250 (Boku Patrol Report No.3 of 1970-71). Table 3.9 gives reported values and volumes of coffee and rice sales through co-operatives 1968-69 to 1970-71.

Cattle

The most recent agricultural innovation in Southwest Bougainville is the introduction of cattle. This is in

Table 3.8

Copra production, by volume and value, Southwest Bougainville, 1968-69 to 1970-71

Society area	Volume (tonnes) 1968-69	1970-71	1971-72	Value of sales ($) 1968-69	1969-70	1970-71
Bana	n.a.	0(30)*	1	1,647	239 (3,600)*	-
Siwai	n.a.	25	36	6,491	2,766	5,105
Buin	n.a.	32	54	5,121	5,891	4,559
Total		57(87)	91	13,259	8,895(12,495)	9,664

* Boku Patrol Report No.3 of 1970-71, p.17 gives volume of production for the Bana area as 30 tonnes for 1970 valued at $3,600.

Sources: Volume: Co-operatives Development Officer, Buin, Annual Report 1971-72, Buin area, pp.2 and 3 (originally in tons).
Value: Co-operatives Division, Department of Business Development, Port Moresby.

Table 3.9

Value of coffee and rice sales through co-operatives, 1968-69 to 1970-71

Society	Coffee 1968-69	1969-70	1970-71	Rice 1968-69	1969-70	1970-71
	$	$	$	$	$	$
Bana	1,001	13	394	-	-	-
Siwai	1,669	667	716	260	192	48
Buin	-	-	259	-	205	-
Total	2,670	680	1,369	260	397	48

Volume of coffee sales (tonnes)

	1970-71	1971-72
Bana	0.41	0.20
Siwai	3.25	0.61
Buin	3.25	1.63
Total	6.91	2.44

Sources: Value: Co-operatives Division, Department of Business Development, Port Moresby.
Volume: Co-operatives Development Officer, Buin, Annual Report 1971-72, Buin area, pp.2 and 3 (originally in tons).

accord with the Australian Administration's policy designed to broaden the agricultural base and reduce Papua New Guinea's dependence on a limited range of tropical crops. In its conception the program was to build up the numbers of cattle on European holdings and government stations, and with the help of Development Bank finance gradually to establish indigenous cattle projects. Since there were originally very few cattle in the country, the initial stock had to be brought in from northern Australia, a procedure which was both costly and time-consuming.

In Southwest Bougainville there were in 1972 two centres of dispersion, the government agricultural station at Konga in Siwai, and the Catholic mission at Tabago, Kono census division. Although both were working towards establishing small-scale cattle enterprises run by individuals or groups of individuals rather than large unified cattle stations in the western sense, there were differences in approach between them. Government policy regarded a holding of about ten heifers and a bull as the minimum economic size of a unit, whereas the mission felt that this was too ambitious and too demanding of the potential owners, and therefore was encouraging individuals or groups to begin with only one or two beasts.

Plate 4. A herd of Brahmin cross cattle on an indigenous cattle project near Munu village, Siwai census division

To qualify for a Development Bank loan (necessary in most cases for potential owners to buy the cattle and provide fencing and yards) a cattle project initiator (an individual or more usually a group of individuals) must be able to provide about 16 ha of land, at least half of which must be in grass, together with good water, fencing and yards before they are supplied with animals. The cost of Brahmin or Brahmin cross animals was about $100-120 each, so that an average loan was about $1,500, $1,200 for stock and about $300 for fencing wire and other materials. Starting with about ten cows and a bull a cattle project operator should be able to sell off some two-year-old heifers or steers while building up his herd. Income from such a project is anticipated to reach $600 in the third year and should increase thereafter. Stocking rates in Southwest Bougainville are about one beast to a hectare. Pastures generally are not very good, but could be improved with fertiliser and good management, so that stocking rates could rise to perhaps 2-3 beasts to the hectare. Most indigenous farmers have little or no experience in cattle, and DASF was endeavouring to send at least one man from each project to a three-month training course at Erap, the main government cattle station in the Markham Valley.

By 1972 there were about thirty cattle projects in Southwest Bougainville, of which about twelve had received stock and were progressing satisfactorily, about seven had not yet received stock, but were proceeding satisfactorily with preparations, and the remainder were small and stagnant. Areally Siwai had about eleven projects mostly clustered within 2-3 miles of Konga, the four southern census divisions had about thirteen, six of which were in villages inland of Tabago, and Nagovisi had only about four. The Buin subdistrict was estimated to have about 350 cattle in 1972, about one-third of the Bougainville total.

Some local people and DASF officers were interested in the possibility of establishing a large-scale co-operative cattle enterprise on some 1,200 hectares of suitable land at present unused and available on the Siwai coastal plain south of Konga. If this enterprise does become established it could be an important breeding centre for the area and also an effective demonstration of relatively large-scale indigenous group enterprise.

The building up of the cattle industry generally in Southwest Bougainville appears an attractive proposition not only because of the relatively good economic returns which potentially could be achieved with the existence of a large potential urban market at Panguna-Arawa-Kieta,[1] but also because the industry requires land (which is available) but relatively much less labour than cocoa growing. In view of a possible future labour shortage this is appealing. Along with other livestock such as pigs and poultry, cattle appear to offer a useful diversification of the agricultural economy which can be achieved at village level. In addition the production of all forms of livestock locally offers the means of providing more animal protein in the diet, again a desirable aim. At present, however, indigenous cattle farmers depend heavily on the knowledge and interest of expatriate agricultural personnel for technical advice, and as yet cattle farming cannot be said to be integrated with the indigenous economy to the degree that it will need to be to survive.

The building of the trans-island road does offer benefits to the industry, primarily through making the movement of cattle into and out of the area much easier.[2] In 1972 no final decision had been made about the location of an abattoir; the two main possibilities being canvassed were a site near Aropa airfield or a site on the Jaba River near the pump station. Whichever was chosen the ability to move cattle to it live by truck would be a great advantage.

A related project is the proposal by BCP to establish water buffalo on the tailings-affected area of northern Nagovisi. This had not yet started in 1972 but the BCP proposal was that the company would establish a trial herd, and if this were successful after two or three years it would assist indigenous farmers to establish some form of water buffalo farming, the details of which had not yet been worked out. The idea is an interesting one, which could, if successful, help provide a solution to the problem of the tailings-affected area.

---

[1] It has been estimated that the total meat consumption of BCP (i.e. at Panguna, Arawa and Kieta) will be 907 tonnes per annum by 1980, of which some 94 tonnes would be pig meats. In addition poultry meat requirements will be 45 tonnes per annum and 100,000 dozen eggs per annum. (DASF, 1971:32; originally in pounds.)

[2] While I was visiting the Konga Agricultural Station a truck load of cattle arrived overland from Kieta. If the road had not existed the shipment would have had to be sent by coastal vessel from Kieta to Buin, unloaded through the surf and shipped by road from Buin to Konga, at a much greater cost and risk of injury or loss.

Fresh fruit and vegetables

The purchase of fresh fruit and vegetables by BCP to supply part of the needs of their mining and construction staffs has provided a major source of cash income to people in the southern part of Kieta subdistrict and to a minor extent in eastern Nagovisi for the last three years. The urban population of Kieta subdistrict, which includes Panguna-Arawa-Kieta, reached a sharp peak of over 16,000 in 1971 at the height of the construction phase of the mine. This figure was expected to fall to about 10,000 in 1972 and to rise gradually thereafter reaching 16,000 in 1979 and 17,500 by 1980 (DASF, 1971:9 and App.V:iv). Such a massive increase created a demand for fresh fruit and vegetables (as well as other foodstuffs and commodities in general) and the company established a buying system for as much produce as could be obtained locally. Existing communications at the time virtually determined that the areas from which fresh foodstuffs could be drawn were on the east side of the island.

In 1970 some 2,469 tonnes of fresh foods worth $226,454 were purchased (DASF, 1971:4). In 1971 the value of purchases rose to $340,697, and although no tonnage figure is available it can be estimated at 1970 values to have been 3,715 tonnes. The value of purchases for January-September 1972 was $145,888, extrapolated to $194,517 for the full year. This would represent about 2,120 tonnes of produce.

The buying of vegetables is organised by sending buying teams to prearranged market points weekly on regular days. These places are often at roadheads to which villagers walk carrying produce which is checked for quality, weighed, and the vendor paid cash immediately. The number and location of these buying points has changed from time to time over the last three years, the major changes being the consolidation of three buying points ('Society', near Rumba Mission in North Nasioi census division, Toborai, and Tokaian) in Arawa town market when this was established early in 1972, and the consolidation of Panguna, Nairovi and Pirurari (which superseded the short-lived Dokotamana buying point) at the same time. Koromira buying point, the most distant from Arawa and Panguna, was closed permanently in November 1971, when demand fell due to the ending of the construction phase. The value of purchases at these buying points for 1970, 1971 and the first nine months of 1972 is shown diagrammatically in Fig.3.8. The peak year in almost all cases was 1971.

The approximate areas served by each buying point are tentatively indicated in Fig.3.9, but these have not been

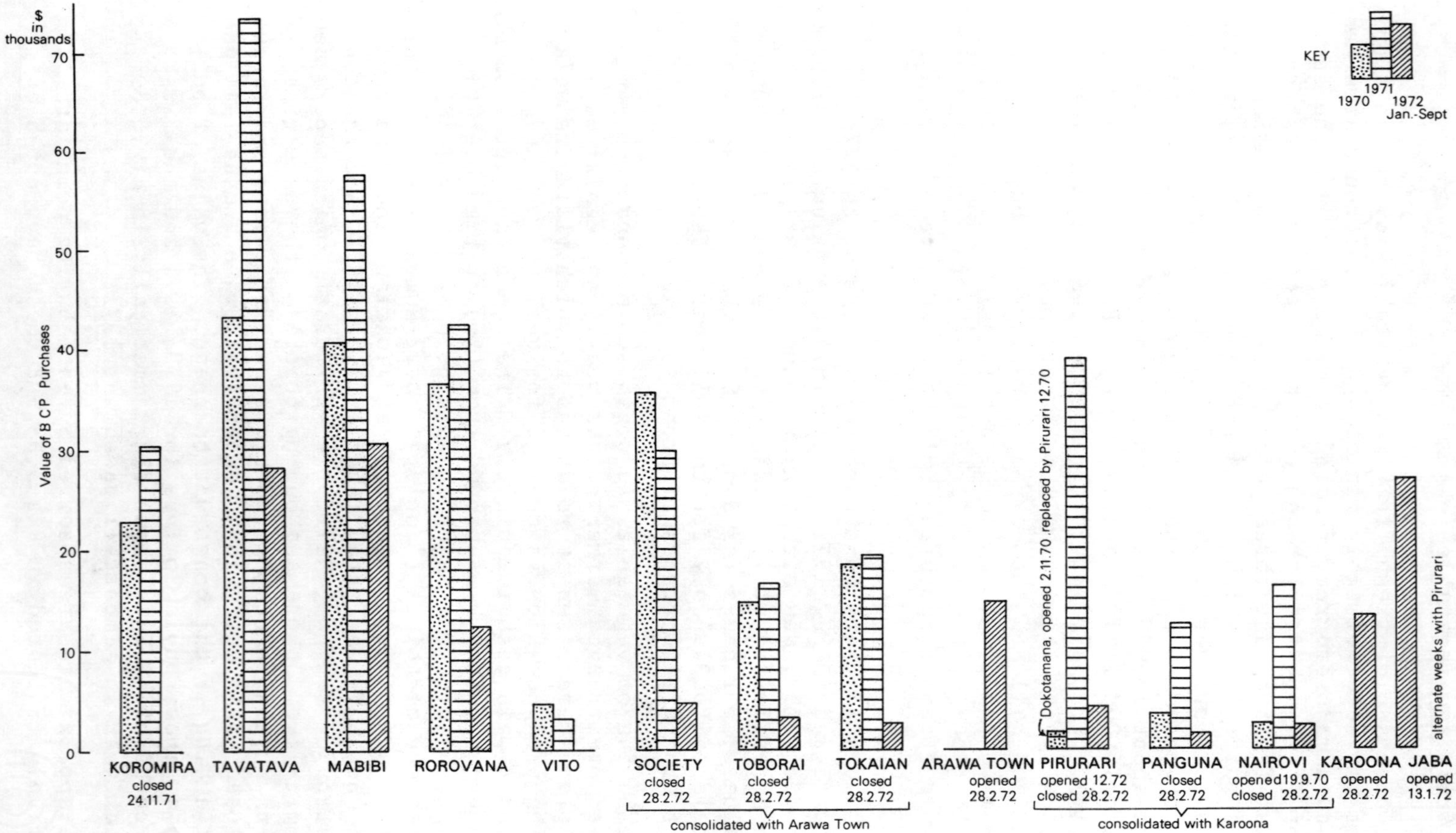

Figure 3.8 Value of fresh fruit and vegetables, purchased by BCP at named buying points, 1970-72

Plate 5. Villagers selling fresh fruit and vegetables to BCP purchasing officers at Arawa market

field-checked. The most significant of these for the present study is that extending into eastern Nagovisi census division centred on a buying point held weekly at Jaba River Pump Station. People from nearby villages originally walked there with produce, but the opening of the trans-island road will have enabled use of vehicles, and thus probably extended the distance from which villagers will find it economically rewarding to bring fresh produce for sale. It is likely that growing fresh vegetables for sale to the Panguna market could become of considerable significance in the Nagovisi census divisions, which appears to have shown less interest than more southern areas in either copra or cattle as alternatives to cocoa to date. However, it may be that the decline in BCP's needs due to the ending of construction will mean that BCP is less interested in buying from this area until its needs build up again in a few years' time. By that time a wholesaling system independent of the Company along the lines suggested in the DASF report (DASF, 1971:28-30) may well have become established as an indigenous enterprise.

A rough calculation of the amount of cash earned per capita by villagers in the Kieta subdistrict by selling vegetables to

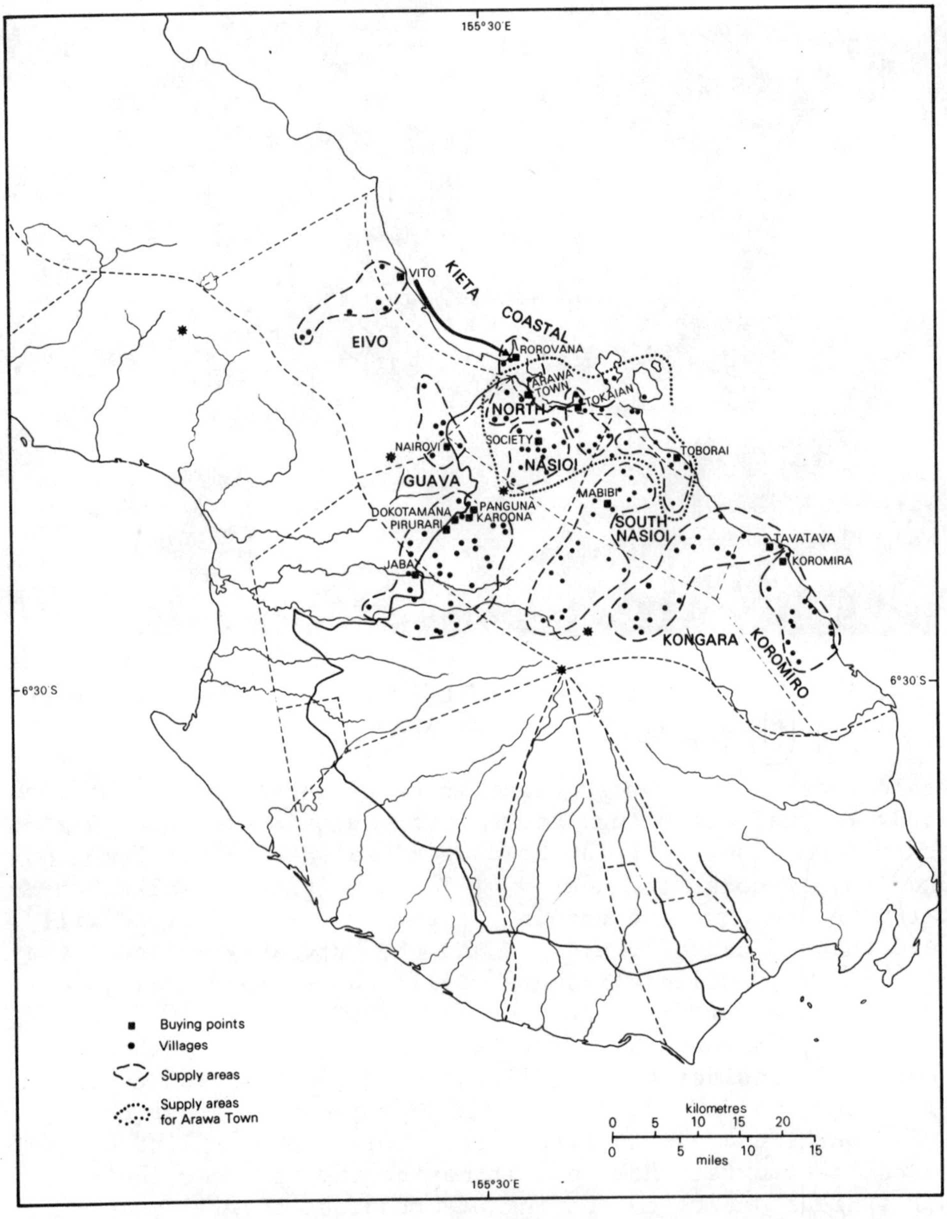

Figure 3.9 Approximate areas supplying BCP buying points for fresh fruit and vegetables

BCP is: for 1970 $14.40 per capita, for 1971 $21.67 per capita, and for 1972 (based on the estimated figure of $194,517 for the full year) $12.37. These calculations are based on 1970-71 village populations (including absentees) for all villages in the subdistrict, but excluding the area of eastern Nagovisi which has been supplying Panguna-Jaba at least in 1971.

The DASF report gives projected demand levels for fresh fruit and vegetables ranging from 3,830 to 5,771 tonnes in 1975 and from 4,917 to 7,813 tonnes by 1980 (DASF, 1971:11-12). Since the returns per man day are higher for fresh fruit and vegetable production than those obtainable from either cocoa or copra, plantation labour, or unskilled urban employment (DASF, 1971:17-19), it seems likely that some villagers at least will want to engage in this economic activity. In the immediate future these are likely to be in Kieta subdistrict, but with the improvement of the Moratona-Jaba Pump Station road, and feeder roads in Nagovisi census division it is likely that villagers in Nagovisi will turn to this form of economic activity. The degree to which Nagovisi will replace Kieta subdistrict as a source of supply, if it does at all, will depend on comparative costs of production and transport between the two areas, as well as on fluctuations in the price of cocoa, and on alternative opportunities for cash earning.

Man-land ratios in Southwest Bougainville

It is widely assumed that Southwest Bougainville as a whole has abundant land available for future use, and this assumption is valid at least for the immediate future. An unpublished study indicates that Bougainville Island has the potential to support about three times its 1966 population (Mount, 1973:10), with a range about that figure according to a variety of assumptions.

Using land capability data supplied by CSIRO, estimates can be made of the capacity of Southwest Bougainville under various land use assumptions. In making these calculations the land use capability map, part of which is reproduced here as Fig.2.7, has been compared with village distribution maps for 1963 (CSIRO) and 1972, and population densities assessed on different classes of land. Inaccuracies are inevitable due to under- or over-enumerations in the population counts, but these are unlikely to be critical in view of the large land surpluses revealed. The areas of land of various classes by census divisions are given in Table 3.10 and the populations supported on them in 1963 and 1972 are given in total for Southwest Bougainville in Table 3.11, and broken down by

Table 3.10

Area of land of various classes by census division
(hectares)

Class	Banoni	Nagovisi	Baitsi	Siwai	Makis	Lugakei	Paubakei	Kono	Southwest Bougainville
IIa	8,032	22,283	8,032	34,980	10,364	5,960	10,623	21,506	121,780
IIc	7,514	-	-	1,166	933	-	1,166	2,332	13,111
IIIc	933	-	-	-	-	-	-	-	933
IVa	7,773	5,182	466	12,178	9,846	-	4,664	16,842	56,951
IVc	6,478	-	-	-	-	-	-	-	6,478
VIc	-	1,814	-	6,478	5,700	1,814	700	16,842	33,348
VIIc	19,692	4,146	-	233	466	1,166	-	4,146	29,849
VIIIa	-	4,923	-	6,478	2,332	4,923	-	12,437	31,093
VIIIb	60,113	7,514	-	4,405	1,814	-	2,591	7,773	84,210
Total	110,535	45,862	8,498	65,918	31,455	13,863	19,744	81,878	377,753

Note: The land capability classes are:

IIa Lands suitable for cultivation with minor limiting factors, over 80% or more of area.
IIc Lands suitable for cultivation with minor limiting factors, over 20-50% of area.
IIIc Lands suitable for cultivation with moderate limiting factors, over 20-50% of area.
IVa Lands suitable for cultivation with major limiting factors, over 80% or more of area.
IVc Lands suitable for cultivation with major limiting factors, over 20-50% of area.
VIc Lands unsuitable for cultivation, but suitable for improved pastures or tree crops with moderate limiting factors, over 20-50% of area.
VIIc Lands suitable for tree crops with severe limiting factors, over 20-50% of area.
VIIIa Lands unsuitable for any commercial use, over 80% or more of area.
VIIIb Lands unsuitable for any commercial use, over 50-80% of area.
Lands of other classes do not occur in Southwest Bougainville.

Source: CSIRO, by courtesy of J. McAlpine (original in acres).

Table 3.11

Total populations supported by various land classes 1964 and 1972, Southwest Bougainville

Land class	Population 1964	Hectares/capita 1964	Population 1972	Hectares/capita 1972
IIa	11,875	10.2	22,535	5.4
IIc	600	21.9	595	22.0
IIIc	-	-	115	8.1
IVa	600	94.9	1,120	50.9
IVc	-	-	-	-
VIc	2,450	13.6	3,105	10.7
VIIc	475	62.8	640	46.6
VIIIa	2,250	13.8	2,730	11.4
VIIIb	250	336.8	665	126.6
Total	18,500	20.4	31,505	12.0

Source: Counts of average village sizes.

census division in Table 3.12. It should be noted that it is assumed that the population of a village is supported on land of the class on which the village itself is located. In the absence of detailed maps delimiting village lands and in view of the scale and limited detail of the land capability maps, this is a reasonable assumption.

Throughout Southwest Bougainville there has been an increase in densities on most land classes between 1963 and 1972. This is to be expected in terms of overall population growth. On Class IIa land (the highest quality land in the area) the density has apparently doubled in the eight-year period. Densities on this class of land are highest in Makis, Nagovisi and Lugakei census divisions. The increase in density has been greatest in Banoni, Makis and Paubakei.

Densities on most other land classes have increased, and in some cases where a class of land supported no people in 1963 it was partly occupied in 1972. However, in a few cases density was less in 1972 than in 1963. This may reflect real shifts in the location of villages or inaccuracies in the village location maps. In gross terms these cases are not of great significance.

To arrive at an approximate estimate of the size of future populations that could be supported, calculations have been made of the capacity of the land areas available by classes to support population under three forms of land use - subsistence agriculture using bush fallowing; copra as a sole cash

Table 3.12

Population densities by land capability class, by census division, 1964 and 1972

Land capability class	Banoni	Nagovisi	Baitsi	Siwai	Makis	Lugakei	Paubakei	Kono	Southwest Bougainville
A. Hectares/capita by census division, 1964									
IIa	32.2	6.2	17.9	9.5	17.3	5.2	16.4	14.8	10.2
IIc	16.7	-	-	23.3	18.7	-	1,166 ha.*	46.6	21.9
IIIc	933 ha.	-	-	-	-	-	-	-	933 ha.
IVa	7,773 ha.	25.9	2,598 ha.	81.2	9,846 ha.	-	4,664 ha.	336.8	94.9
IVc	6,478 ha.	-	-	-	-	-	-	-	6,478 ha.
VIc	-	1,814 ha.	-	10.0	6.7	1,814 ha.	670 ha.	17.7	13.6
VIIc	78.8	18.4	-	233 ha.	466 ha.	1,166 ha.	-	4,146 ha.	62.8
VIIIa	-	7.0	-	25.9	15.6	9.8	-	-	13.8
VIIIb	240.5	7,514 ha.	-	4,405 ha.	1,814 ha.	-	2,591 ha.	7,773 ha.	336.8
B. Hectares/capita by census division, 1972									
IIa	11.3	4.1	8.9	5.3	3.9	4.4	5.6	7.0	5.4
IIc	17.3	-	-	10.1	20.7	-	1,166 ha.	2,332 ha.	22.0
IIIc	8.1	-	-	-	-	-	-	-	8.1
IVa	68.8	25.3	2,598 ha.	105.9	48.0	-	18.7	73.2	50.9
IVc	6,478 ha.	-	-	-	-	-	-	-	6,478 ha.
VIc	-	5.2	-	5.7	13.1	2.5	670 ha.	38.7	10.7
VIIc	45.3	20.2	-	233 ha.	466 ha.	1,160 ha.	-	4,146 ha.	46.6
VIIIa	-	9.0	-	18.8	20.3	4.7	-	18.7	11.4
VIIIb	90.4	7,514 ha.	-	4,405 ha.	1,814 ha.	-	2,591 ha.	7,773 ha.	126.6

* Where land of a particular class is present but no villages are sited on it, the acreage of land is given.

crop; and cocoa as sole cash crop. Clearly these are idealised conditions.

The population which might be supported under subsistence bush fallow alone is the most difficult to estimate, since there are no accurate data on planting and fallow areas on the various classes of land. McAlpine (CSIRO, 1967:161) found that overall 0.16 hectares per head per annum was under garden in 1963. At that time there was relatively little land under productive commercial crops. A nation-wide survey of indigenous agriculture about 1961-62 found a national average of 0.13 hectares of planted garden per person during the survey years (TPNG, Bureau of Statistics, 1963:14). If a 10-year fallow period is allowed, McAlpine's figures indicate that 1.62 hectares per capita were used under bush fallow agriculture in Bougainville in 1964. However, if there was no obvious land deterioration the safe density might be much higher. Conversely the poorer classes of land might only support lower densities under bush fallow. In that case a general figure of 1.62 hectares might give an unduly high potential population.

Three arbitrary scales were applied to the amounts of land in each class to give low, medium and high totals for the possible populations which might be supported (Table 3.13).

Table 3.13

Scales for population support by subsistence agriculture, by land capability class

Land class	No. of hectares/capita required		
	Low population	Medium population	High population*
IIa	1.6	1.6	1.0
IIc	3.6	2.0	1.2
IIIc	5.7	2.4	1.4
IVa	7.7	2.8	1.6
IVc	9.7	3.2	1.8
VIc	11.7	3.6	2.0
VIIc	13.8	4.1	2.2
VIIIa	15.8	4.5	2.4
VIIIb	17.8	4.9	2.6

* Implies a 6-year rather than 10-year rotation cycle.

These scales are conservative as they assume no improvements in technology, and, in the low population case, allow for large land areas per capita. Applying these scales to the amount of available land in each class gives estimates of possible populations of 99,000 (low), 145,000 (medium) and 247,000 (high). No allowance has been made for the amount of land used for roads and settlements, but this is unlikely to be more than one per cent of all land available. Although these figures are only rough approximations, the important point is that they are conservative. Yet even the low density assumptions show that a bush fallow system of agriculture could support three times the present population under a subsistence economy, with no changes in technology.

Table 3.14 gives a breakdown by census division for the low estimate.

Table 3.14

Low estimates of population supportable by bush fallow agriculture

Census division	Low estimate	1972 population
Banoni	13,671	2,535
Nagovisi	15,623	6,825
Baitsi	5,021	850
Siwai	24,729	8,559
Makis	8,706	3,401
Lugakei	4,231	2,798
Paubakei	7,691	2,014
Kono	19,068	4,088
Total	98,740	31,070

In estimating the population which might be supported if all suitable land were used for cocoa production, a number of assumptions must be made, including the following: that only land of classes IIa, IIIc, IVa and IVc is regarded as suitable for cocoa production; that the size of a cocoa farm necessary to support a family is 4.0 hectares on class II land, 6.1 hectares on class III land, and 8.1 hectares on class IV land; that the average family size in Southwest Bougainville is 6.5 persons; that the proportion of classes II, III and IV land over 2,000 feet (above this altitude cocoa is not grown) is the same for Southwest Bougainville as for the whole island, namely 4 per cent (Mount, 1973:4); that the proportion of inaccessible land of classes II, III and IV in Southwest Bougainville is the same as for the island as a whole, namely 18 per cent (Mount, loc. cit.); that no allowance is made for non-agricultural land requirements; and that the return from cocoa growing remains at least as favourable as at present.

Ignoring for the moment the limitations of altitude and accessibility, the number of farm units per census division which could be accommodated under these assumptions is given in Table 3.15. If 22 per cent is deducted to allow for limitations of altitude and accessibility, the number of people which could be supported is 193,000. Table 3.16 gives a breakdown of their distribution by census division.

Table 3.15

Possible number of cocoa farms by census division

Census division	Total
Banoni	3,898
Nagovisi	6,144
Baitsi	2,042
Siwai	10,144
Makis	3,776
Lugakei	1,472
Paubakei	3,200
Kono	7,392
Total	38,068

Table 3.16

Population theoretically supportable by cocoa by census division

Census division	Population	1972 population
Banoni	19,763	2,535
Nagovisi	31,150	6,825
Baitsi	10,353	850
Siwai	51,430	8,559
Makis	19,144	3,401
Lugakei	7,463	2,798
Paubakei	16,224	2.014
Kono	37,376	4,088
Total	192,903	31,070

The third main alternative land use in Southwest Bougainville is copra production. The case that all suitable land would be used solely for copra production is an unlikely one, but it is examined here. The assumptions made include: that land of classes IIa, IIc, IIIc, IVa and IVc is regarded as suitable for copra production; the size of farm necessary to support a family is 6.1 hectares on class II land, 8.1 hectares on class III land and 10.1 hectares on class IV land; average family size is 6.5 persons; that 22 per cent of the land is too high or inaccessible for copra production; and that yields of copra are equivalent to the national average (on which farm size estimates are based). This last is not at present the case in Southwest Bougainville. If these assumptions are met the population which could be supported is given by census division in Table 3.17.

Table 3.17

Population theoretically supportable by sole copra, by census division

Census division	Supportable population
Banoni	20,701
Nagovisi	21,198
Baitsi	6,941
Siwai	36,276
Makis	14,368
Lugakei	4,974
Paubakei	12,178
Kono	28,336
Total	144,972

In fact, on national averages a mix of cocoa and coconuts produces a higher return to smallholders than either sole cocoa or sole copra production (Emmery, 1970:Table 8) but this may not be so in Southwest Bougainville at present, due to the low yields obtained from coconut palms in this area. However, it remains possible that this crop combination could support a higher population than could sole cocoa. A similar set of calculations to those already given was completed using the same assumptions as for sole cocoa, except for assumed farm sizes of 3.2 hectares on class II land, 4.9 hectares on class III land and 6.5 hectares on class IV land. This produced the results given in Table 3.18.

If the assumptions used in making these calculations are reasonable it would seem that Southwest Bougainville could certainly support a population several times greater than at present. These figures should, however, be taken only as broad indicators and not in any sense as goals to be attained. Nor should it be inferred that no 'population problem' exists. Access to land is only one of many requirements for a growing population and the difficulties of maintaining, let alone expanding, educational, medical and other services for a

Table 3.18

Population theoretically supportable by joint cocoa/copra production, by census division

Census division	Supportable population
Banoni	24,700
Nagovisi	38,935
Baitsi	12,941
Siwai	64,285
Makis	23,933
Lugakei	9,328
Paubakei	20,280
Kono	46,845
Total	241,247

population growing at 4 per cent per annum are severe. The simple calculations of carrying capacity made above assume equal access to the land by those who need it. This is unrealistic. Demographic processes will ensure that some groups will increase in size while others decline. Land ownership rarely changes in accord with such moves now that traditional means of adjustment (such as warfare) are proscribed. Furthermore, the local migration which would be necessary to ensure full utilisation of usable land is unlikely to occur without some stresses. It will never be completely in phase with changes in land availability. Finally the process of urbanisation and expansion of the secondary and tertiary sectors will undoubtedly have a great impact on population distribution.

In the wider context it can be argued that Southwest Bougainville will have a population problem in the foreseeable future.[1] If the present rate of growth were to continue, the population of the area would reach 100,000, three times its present size, in thirty-two years. It will double again in a further seventeen years. The costs of such rates of increase, together with the relative shortage of working age population which high rates of increase imply, suggest the need for measures to assist people to reduce the birth rate when they wish to do so, and for considering the supplementation of the workforce through adult immigration. Active measures to

---

[1] In the short term it may have a different population problem - that of labour shortage (see Chapter 4).

diversify the local economy by expansion of secondary and tertiary activities would also be desirable, both as a means of relieving local pressures on rural land, and as a means of bringing rural areas into closer links with the modern sector.

Consideration of the distribution of potentially usable land, and the existing variations in population density in relation to this land suggests that if cocoa or other cash crop planting continues to expand as rapidly as in the recent past, direct population pressure on land will first become evident in the census divisions of Nagovisi, Siwai, Makis and Lugakei. This should be taken into consideration in any plans for the introduction of family planning programs, or when locations are sought for the establishment of secondary or tertiary service centres. If sites for cocoa growing resettlement schemes are sought, then the census divisions of Baitsi, Paubakei, Kono and Banoni, with their greater areas of unused land, should be examined first.

If cocoa planting and production continue to expand rapidly, the demand for tertiary services, and opportunities for local involvement in trucking, processing, marketing and retailing will also grow. Integrated planning on an area basis would help make the most of the opportunities afforded by the rapid cocoa development already experienced in Southwest Bougainville.

Non-agricultural economic activities

Basket making. A craft industry of continuing minor importance in the southern two-thirds of the area is the making of fine baskets from a plant of the *Lygodium* genus. The baskets, described as 'Buka' baskets outside Bougainville, are made mostly in Siwai and the area near Buin, and command a ready sale to Europeans, both in Bougainville and outside. The influx of relatively wealthy Europeans to Panguna-Arawa-Kieta, willing to pay high prices, disturbed the market about 1970. This resulted in the production of large numbers of imperfect articles and overbuying by the Buin and Siwai Producers' Co-operatives which handle most of the sales. This situation will be corrected by greater control of quality and marketing by the co-operatives. Table 3.19 gives the value of basket sales by co-operatives for three recent years.

Cash returns from basket making in the Siwai census division were estimated to be about $30,000 per annum in 1972, or $25 per nuclear family (Rural Development Officer, Siwai, Annual Programme and Estimates, 8.9.72).

Table 3.19

Value of basket sales by co-operatives, 1968-69 to 1970-71

Co-operative	1968-69	1969-70	1970-71
	$	$	$
Bana	34	2	18
Siwai	10,692	12,870*	33,224
Buin	9,341	9,389	5,114

* Total turnover from baskets for year 1969-70 was $18,961 (Konga Patrol Report, No.1 of 1970-71, App.5).

Source: Co-operatives Section, Department of Business Development, Port Moresby.

Plate 6. Patterned Siwai mats, baskets and tray

This local industry, producing a unique article for which there is a demand both within and without Papua New Guinea deserves to be encouraged and developed, particularly by the introduction of new styles and in the marketing of the products.

The skill required appears to be widely known, particularly amongst Siwai men (Oliver, 1955:14). The manufacture of a small basket selling for about $10 is reported to take about two days, hence it is a useful way of raising ready cash.

Trade stores

In 1972 there were about 265 trade stores in Southwest Bougainville. Their distribution is shown in Fig.3.10 and a detailed village list given in Appendix III. This averages slightly over one per village, but not every village has a trade store while some have several, Kanauro village in Paubakei census division having as many as eight for a population of 255. Table 3.20 lists the population per trade store by Local Government Council areas.

Table 3.20

Trade stores and populations, 1972

L.G.C. area	No. of trade stores, 1972	Population 1972	People/trade store
Bana	58	10,210	176
Siwai	68	8,559	126
Buin	139	12,301	88
Southwest Bougainville	265	31,070	117

Source: Trade store data obtained from Local Government Councils.

Trade stores are most numerous in the Buin Local Government Council area where there are very few villages without at least one store, even in the most inaccessible inland areas. Most villages in Siwai Local Government Council area also have at least one store or if not they are all, except Iru village, within a short distance of villages which do. The picture for the Bana area is obscured by the fact that locations were not available for seven stores and the locations of three others could not be identified. However, the majority of these are likely to have been located in Nagovisi or Baitsi census divisions, particularly in Beretemba, Boku, Lopere, Lotari, Mosigetta, Pikei and Agabai villages. Most villagers in Nagovisi and Baitsi would find a trade store in a neighbouring village if there was not one in their own. In contrast there are very few trade stores in Banoni census division and most of those which exist are in the southern part.

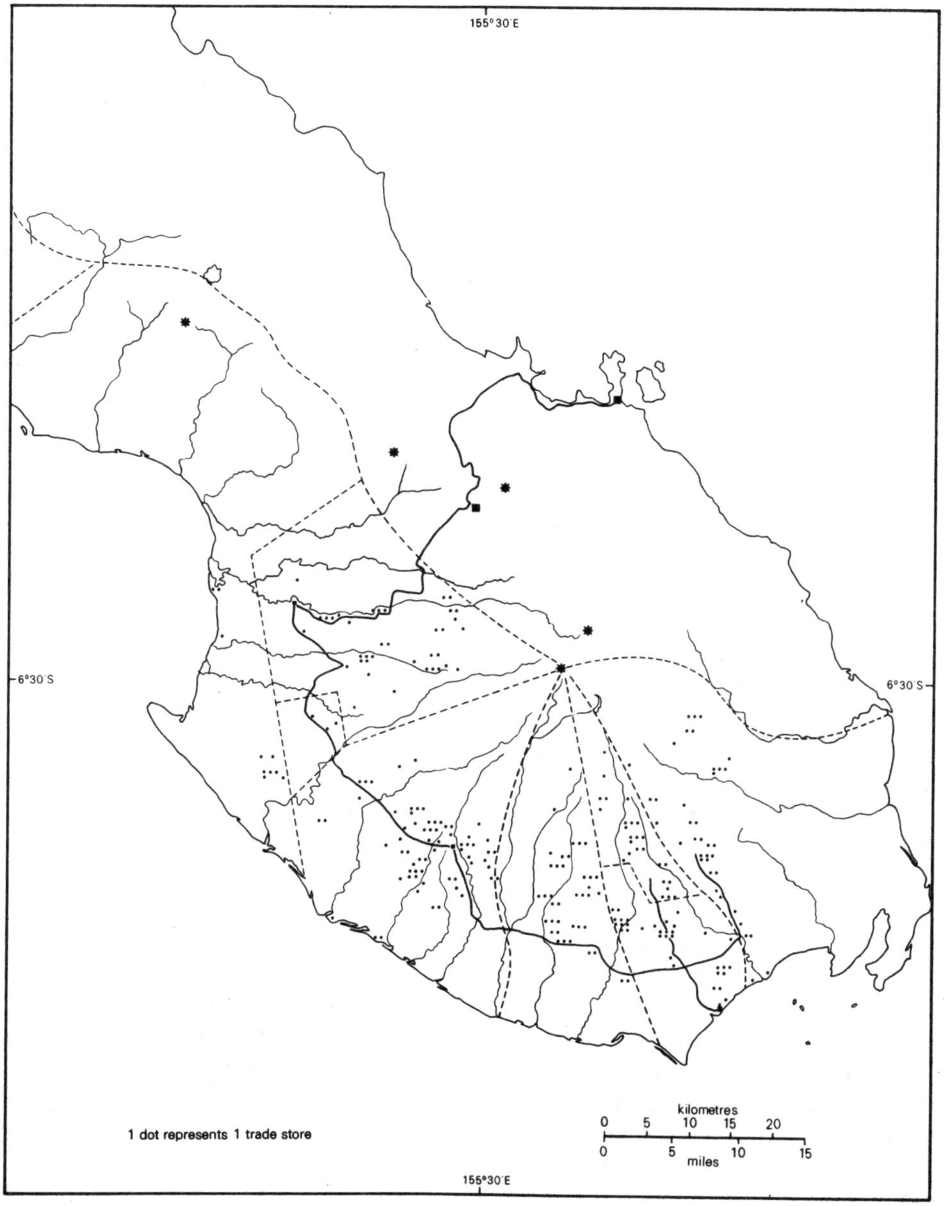

Figure 3.10 Distribution of trade stores, 1972

The presence of so many stores reflects the impact of the cocoa economy, which provides cash for food and drink, clothes, radios, household utensils and other store goods. There are no very reliable earlier data for comparison but a 1968 survey recorded 151 trade stores in Buin subdistrict in that year (Office of the Economic Adviser, Dept of Trade and Industry, [c.1969]:Table 1.19,3).

It is well known that the turnover of the majority of trade stores in Papua New Guinea is small. No data were collected on this topic in 1972, but the 1968 survey gives a total annual sales figure of $31,596 and an average annual sales figure of $209 for Buin subdistrict (loc. cit.). The average for Bougainville District in that year was given as $393, and the Southwest was well below the east ($4,931) and north ($793) of the island (loc. cit.). It is likely that average sales per store will be higher in Southwest Bougainville with increasing cash income, but there are still likely to be many very small stores which earn their owners prestige but little income.

The goods sold in trade stores have to be imported into the area. Many but not all trade store owners belong to one of the three wholesalers' co-operatives: Telei Co-operative Consumers Ltd (Buin), Motuna Wholesale Co-operative Ltd (Siwai) and Tutuna Co-operative Ltd (Bana area). Some indigenous trade store owners have large enough businesses to be able to undertake their own indenting from suppliers in Kieta. For smaller operators the wholesaling co-operatives provide relatively efficient wholesaling service, and it has been claimed that these co-operatives have reduced wholesale prices by at least 15 per cent on most items compared with the prices of private suppliers (pers. comm., Co-operatives officer, Buin, October 1972). In addition to acting as wholesalers for indigenous trade store owners the wholesale co-operatives operate their own retail stores in the main centres. The annual turnover for the Telei Co-operative Society rose from $44,636 in 1968-69, through $54,885 in 1969-70 to $121,326 in 1970-71. For the Tutuna Co-operative Society the annual turnover was $101,346 in 1970-71 (Co-operatives Section, Dept of Business Development, Port Moresby). Despite earlier problems with bad debts, the turnover of Telei society shows a healthy increase over this period. The indications are that all three societies should prosper in future, provided they can obtain staff and management of adequate calibre to run their businesses, and maintain the confidence of the people.

All three societies, but particularly the two northern ones (Tutuna and Motuna) stand to benefit by the reduction of transport costs brought about by the opening of the trans-island road. The consumer goods brought in by the wholesaling societies form the backload against the outward shipment of cocoa. Of these goods rice is probably the most important. In 1972 as a result of the change to direct indenting from Kieta by the wholesale co-operatives and the change to road rather than sea transport the price of a 56 lb (25 kg) bag of rice at Konga dropped from $6.30 to $5.10 (pers. comm., Co-operatives Officer, Buin). Table 3.21 gives a breakdown of the costs of shipping rice from Kieta to Beretemba, Konga and Buin by sea and by road.

Table 3.21

Comparative sea-road freight costs on rice, Kieta-Konga
($/ton)

Elements of costs	To Beretemba	To Konga	To Buin
By sea			
Freight Kieta-Kangu	12.00	12.00	12.00
Stevedoring at Kangu	2.40	2.40	2.50
Road freight Kangu-destination*	17.00	10.00	1.50
Total cost	31.40	24.40	15.90
By road			
Freight Kieta-destination	12.00	12.00	17.00
Saving /ton	+19.40	+12.40	-1.10
/56 lb bag	+48.5 cents	+31 cents	-2.75 cents

* Using Couper and Cotter's one way road haulage charges (Couper and Cotter, 1972:68).

Total rice imports to Southwest Bougainville in 1972 were likely to have been about either 400 tons (see Table 7.5) or 300 tons (Table 7.6). Assuming that this was equally divided between the three parts of the area the transport cost savings on rice theoretically attributable directly to the new road (if it had been open for the full year) would have been $1,940 to Beretemba, plus $1,240 to Konga less $110 to Buin (if its share had been shipped by road), i.e. $3,070 in all for the lower figure and $4,083 for the higher. This example demonstrates the economic importance of the new road to the wholesaling or consumers' co-operatives, and through them to trade store owners and the population who are their customers. However, the road may also have a deleterious effect on the smaller and less economic trade stores, which may lose custom to the larger stores, or to the now more readily accessible supermarkets and stores in Panguna, Arawa and Kieta.

Tourism and related service industries

These forms of economic activity are at present minimal in Southwest Bougainville, and so far there is little indigenous participation. There is, however, some potential for further development. Existing facilities for visitors in the area are limited to two small expatriate-owned guest houses in Buin town, and there are no cafes, restaurants or motor vehicle service stations, though food and drink and petrol may be purchased at some trade stores, either expatriate or indigenous owned. One enterprising expatriate business had responded to the opening of the new road by running motor 'safaris' by 4-wheel drive vehicle between Kieta and Buin. Two of these had taken place (with some difficulty) by September 1972 when I visited the area, but it was too early to say whether the enterprise would continue.

The physical resources of Southwest Bougainville do not readily encourage the idea of tourist development; there are no attractive beaches or coral areas as there are in the east and north and the climate is much wetter than in the east and north. One can imagine a few intrepid visitors undertaking the arduous climb to Lake Loloru and the volcanic peaks of Mts Taroka and Takuan, but without the religious significance of a Fujiyama and given the almost constant cloud and rain of the higher altitudes these features can hardly be counted major attractions. However, the history of the area as a battleground of World War II, and the presence of some relics of this period are a potential attraction to Japanese and Australian visitors. If the people of Southwest Bougainville wanted to encourage visitors (and they may not) much could be done to use the cultural resources of the area as a drawcard. These might include the setting up of cultural centres where people could demonstrate traditional skills such as house building, basket making, string bag netting, pottery making, traditional weapons, _garamut_ making and sounding, singing and dancing, garden techniques and many others. Artists and craftsmen in the modern as well as the traditional styles could be encouraged to establish studios, perhaps at the L.G.C. centres. There is considerable scope for indigenous business enterprises such as guest houses or hostels, cafes and restaurants, and vehicle service stations. The opening of the new road has already encouraged indigenous and expatriate visitors to come into the area at weekends, and this is likely to continue as the quality of the road is improved.

The role of the Development Bank

The development of commercial enterprises by indigenous people in Southwest Bougainville has been assisted by the Papua New Guinea Development Bank to an increasing degree in recent years and this support is likely to increase, particularly now that the trans-island road has linked the area to Kieta. In 1971-72 the Bank's investment of capital in this area totalled over $200,000, some 26 per cent of its investment in Bougainville District. Table 3.22 gives a breakdown of this investment by type of economic activity for 1971-72 and part of 1972-73. The greatest number of loans were for trucks and tractors (averaging $2,333 each), cattle projects (averaging $1,458) and trade stores (averaging $900), while the greatest share of money went into property developments in Arawa town centre, by the three producer societies. In these cases the land has been bought and the building of several retail shops will be undertaken by the societies; the shops will be leased to tenants. The Bank hopes that its investment in trucks will help provide custom for these shops via the new trans-island road.

Table 3.22

Development Bank investment in Southwest Bougainville, by type of enterprise, 1971-72 and 1972-73 (incomplete)

Type of enterprise	No. of loans 1971-72	Amount 1971-72 ($)	No. of loans 1972-73	Amount 1972-73 ($) (incomplete)
Carrying and cartage	23	51,351.50	2	7,237
Shipping and boats	1	113		
Retail trade	6	5,400		
Property	3	113,000		
Building and construction	2	10,066		
Beef cattle	14	20,150	1	1,720
Poultry	1	300		
Cocoa	4	772	1	520
Coconuts	1	525		
Fishing	1	317		
Engineering			1	500
	56	201,994.50	5	9,977
Total for Bougainville	145	766,213		

Source: Development Bank, Port Moresby.

The location of the Development Bank's investments is shown, so far as it can be determined, in Fig.3.11. A breakdown of the total investment by L.G.C. areas within Southwest Bougainville is given in Table 3.23.

Table 3.23

Development Bank investment by L.G.C. area

L.G.C. area	Amount, 1971-72 and part of 1972-73 ($)	Per cent	$/per capita
Bana	58,340.60	27.6	5.71
Siwai	67,713.40	32.0	7.91
Buin	85,600.50	40.4	6.96
	211,654.50	100.0	6.81
Unlocated	317		
	211,971.50		

This analysis shows that the northern third of the area has received rather less than one-third of the total investment and less than the average per capita. The Siwai area has received about one-third of the investment and more than the average per capita, while the southern third has received more than one-third of the total, and very slightly more than average per capita. In fact all the investment in the northern area went to Nagovisi census division unless the cover addresses of Moratona mission and Boku patrol post disguise locations of recipients in Banoni and Baitsi census divisions. This would seem to indicate that the northern third might in future receive rather more investment than the southern two-thirds, particularly as the opportunities for business development generated by the new road will be felt more strongly there than further south.

The availability of capital from the Development Bank, although relatively small in total, is important in the development of indigenous business enterprise. Only 5 per cent of the total (two loans to an earth-moving business located in Buin town) was made to an enterprise in which expatriates were involved, and in this case there are also indigenous partners in the concern and the work of the firm is largely concerned with the improvement of feeder roads throughout the whole area. Apart from the practical benefits derived from the availability of capital the psychological effects of the Bank's willingness to lend to indigenous entrepreneurs must be extremely valuable.

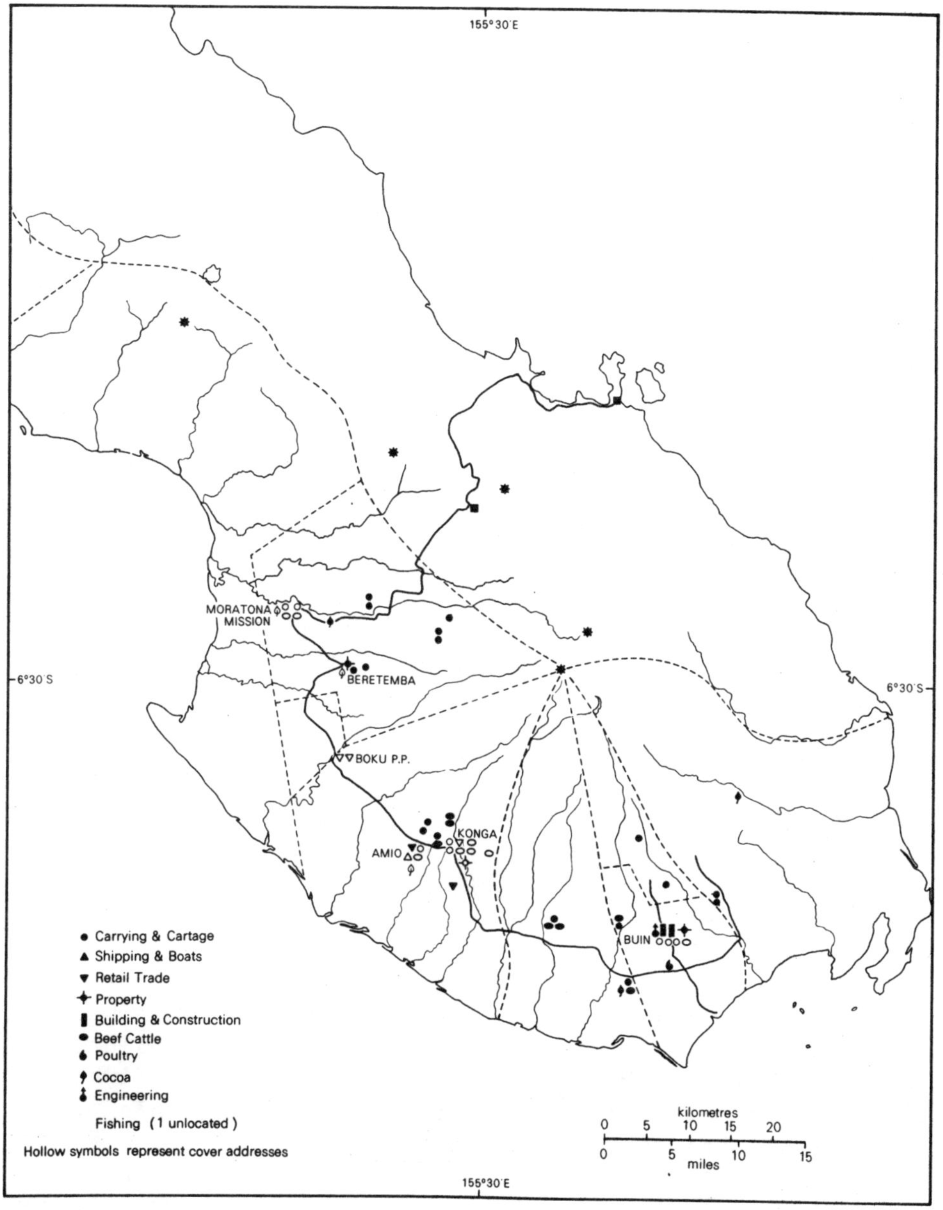

Figure 3.11 Development bank loans, 1972-73

Present incomes from agriculture and baskets

An attempt has been made to assess regional and per capita incomes from agriculture and basket making for Southwest Bougainville in 1972. Four sources of income only have been included: sales of cocoa, copra, baskets and fresh fruit and vegetables. Income from sales of coffee and rice has been excluded since they are of negligible importance compared with cocoa, and the data are lacking. Cattle projects do not yet provide significant incomes in the area. Incomes from retail trading are excluded because data are lacking, and also because the money spent at trade stores has already entered the area from the outside sale of agricultural produce and to count it again would be double counting.

Regional income from cocoa and copra has been calculated by multiplying the tons of cocoa and copra produced in 1971-72 in each of the three L.G.C. areas (Bana, Siwai and Buin) by the 1972 prices of cocoa and copra less estimated transport costs by sea through Kangu to Kieta. Thus for cocoa the price at August 1972 was $490/ton (Co-operatives File BD 12-4-1, Buin, Co-operatives Monthly Reports), estimated transport costs were for Bana $32/ton, Siwai $24/ton, Buin $20/ton; hence the multipliers used were $458, $466 and $470 respectively. Similarly the Copra Marketing Board's price of copra at Kieta in 1972 was $140/ton, hence $108, $116 and $120 were used as multipliers for copra. Basket sales for Siwai and Buin are estimated figures. Fruit and vegetable sales for twelve villages in eastern Nagovisi are calculated for 50 weeks at an estimated figure of $350/week (Boku Patrol Report, No.3 of 1970-71) and are well within the total disbursements by BCP at the Panguna-Jaba market complex for 1972. The results of the calculations are given in Table 3.24. These figures indicate a regional income from agricultural sources of some $650,000 for 1971-72. On a per capita basis Siwai and Buin areas show a higher figure than Nagovisi, which in this year was handicapped by the highest transport charges within the area. The indications are, however, that the Nagovisi area will show the greatest future expansion in cocoa production and also will benefit most from reduced transport costs brought about by the new road. Hence its income per capita can be expected to rise more rapidly than that of Siwai and Buin in the next few years.

Nagovisi is rather better off than the above figures show since people in that area have received cash compensation payments from BCP for the loss of land, loss of fishing, and as occupation fees. These have been estimated to amount to

Table 3.24

Incomes from agriculture, Southwest Bougainville, c. 1971-72

Source	Bana	Siwai	Buin	Southwest Bougainville
	$	$	$	$
Cocoa sales	130,988	178,944	276,360	586,292
Copra sales	108	4,060	6,360	10,528
Basket sales	-	30,000	5,000	35,000
Fruit and vegetable sales	17,500	-	-	17,500
Regional income	148,596	213,004	287,720	649,320
Income per capita	14.55	24.89	23.39	20.90

Source: see text.

$134,000 in total to February 1972. Future annual cash payments to the Nagovisi area are estimated to be about $97,500 (W.D. Scott and Co., 1973:3-10/3-11). Hence if a figure of $100,000 is added to those in Table 3.24 for approximate compensation received in 1971-72 the regional income for the Bana area rises to $248,596 and the per capita income to $24.35. Regional income for Southwest Bougainville from agricultural sources and compensation payments rises to $749,320 or $24.12 per capita.

A final important additional source of income in the area is that earned by individuals employed outside the area, or within it. This topic is discussed in the next chapter.

Chapter 4

Labour and employment

In most parts of Papua New Guinea a sizeable proportion of the *de jure* village population now lives away from its 'home' villages (R.G. Ward, 1971:94). In the Banoni, Nagovisi and Siwai census divisions of Southwest Bougainville the proportion of all absentees is between 20 and 30 per cent, and totals of a similar order are probably true also for the southern census divisions. About two-thirds of these absentees are students studying either within Bougainville or outside it. For Southwest Bougainville the majority of students are attending schools within the district and hence probably return to their villages at frequent intervals. Most of the remaining absentees are adult males, and again the majority are still within the district, though not necessarily in Southwest Bougainville. A few hundred adults from Southwest Bougainville are working outside Bougainville, the Rabaul area being a favoured location although they are to be found widely scattered throughout Papua New Guinea.

The available information on numbers of people absent from villages in Southwest Bougainville in 1972 is given in Table 4.1. Unfortunately the data for the northern and southern parts of the area are not comparable. For Makis, Lugakei, Paubakei and Kono census divisions only absentees 'Resident outside Electorate' are available. This presumably refers to the South Bougainville electorate which coincides with Buin subdistrict. It is likely that the numbers of both children and adults absent from villages in the four southern census divisions but within the district would be comparable to those for the northern census divisions, which would raise the southern census division totals considerably.

These figures do not show any very strong subregional differences, though the fact that Siwai census division has the highest proportion of adult absentees may reflect somewhat higher pressures of population on land. It is possible that the high proportion of all absentees in Banoni census division reflects the lack of roads, the relatively small development

Table 4.1

Numbers absent from villages in 1972, by census division

Census division	Children*		Adults**		Total absentees	All absentees as % of village population	Adult absentees** as % of village adult population
	M	F	M	F			
Banoni***	287	247	194	24	752	29.7	19.0
Nagovisi***	582	524	418	48	1,572	23.0	14.8
Baitsi***	15	21	64	8	108	12.7	16.4
Siwai***	786	705	652	126	2,269	26.5	20.2
Subtotal***	1,670	1,497	1,328	206	4,701	25.0	17.8
Makis≠	14	4	220	26	264	7.8	14.3
Lugakei≠	19	6	154	33	212	7.6	12.9
Paubakei≠	9	11	97	23	140	7.0	11.5
Kono≠	21	12	185	65	283	6.9	11.8
Subtotal≠	63	33	656	147	899	7.3	12.7

* Includes 18 'child workers', presumably the children of absent workers.
** Includes adult students.
*** Includes all absentees whether inside or outside the district.
≠ Includes only those 'Resident outside Electorate'.

Source: Division of District Administration census records.

of cocoa and the lack of alternative employment opportunities in that census division. The converse would be true for Baitsi census division.

Generally the percentage of adults absent from their villages is not excessively high. This reflects the importance of cocoa as a cash crop, and possibly the rather self-contained nature of Southwest Bougainville society.

In an attempt to determine whether there have been significant changes in the number of absentees in the last five or seven years a comparison has been made between numbers of adult worker absentees in 1972 and at the date of the previous census. The results are given in Table 4.2. In Banoni census division there has been a 4 per cent decline in the proportion of absent adult male workers, but a slight rise in the proportion of absent male and female workers between 1967 and 1972.

Table 4.2

Changes in adult absentee percentages, 1965-67 and 1972

Census division	Adult male worker absentees as percentage of all adult males		Adult male and female worker absentees as percentage of all adults	
	1965-67	1972	1965-67	1972
Banoni	26.6 (1967)	22.3	16.4	17.7
Nagovisi	34.7 (1967)	29.6	13.0	13.0
Baitsi	22.9 (1967)	22.4	21.3	12.3
Siwai	25.0 (1965)	29.4	16.7	18.2
Makis	38.1 (1965)	22.6*	22.8	14.3*
Lugakei	19.1 (1966)	19.0*	11.1	12.9*
Paubakei	24.3 (1965)	16.7*	15.0	11.5*
Kono	45.6 (1965)	16.2*	25.5	11.8*

* Includes only those 'Resident outside Electorate', and hence not comparable with 1965-67 figures.

Source: Village censuses for named years.

In Nagovisi there has been a decline of 5 per cent in the proportion of absent adult male workers but no change in the proportion of absent male and female workers. In Baitsi there has been little change in the proportion of absent male workers, but a 9 per cent fall in the proportion of absent male and female workers. In contrast Siwai census division shows a 4 per cent rise in the proportion of absent male workers and a 2 per cent rise in the proportion of absent male and female workers. The figures for the four southern census divisions cannot be compared between censuses, but in 1965 Kono and Makis census divisions had relatively high proportions of both male workers and male and female workers absent, while Lugakei had the lowest proportions absent of both, out of all eight census divisions.

As with the 1972 village distributions of absentees there appears to be little readily explainable pattern in these differences, although the high 1965 values for Kono census division probably reflect its relative lack of access and job opportunities. Nagovisi, and Baitsi and Banoni (to a lesser degree), and Makis, Lugakei and Paubakei and possibly Kono census divisions all experienced rapid expansion of cocoa planting during the period 1965-72 (see Tables 3.1 and 3.2), which may explain the decline in the numbers of absent male workers, while Siwai experienced a much less rapid expansion of cocoa during that period, and this may account for the relative increase in the proportion of male and all adult workers absent. In view of the inherent inaccuracies of the census taking techniques, and the reduction in frequency of counts in the later 1960s, it may be that the differences in rates between the two periods is largely a result of enumeration errors

The proportions of absent male workers in Table 4.2 are, with the exception of Kono and Makis census divisions, at the lower end of the national range as demonstrated by R.G. Ward (1971:94). Several census divisions on the Papuan coast, Morobe District, and northeast Bougainville had over 50 per cent of the adult male population absent in 1967-69.

Figure 4.1 shows the distribution of absent adults expressed as a percentage of total village population in 1972; again the data for the four southern census divisions refer only to those absent outside the electorate. There appears to be no significant correlation of adult absentees with the degree of accessibility of villages, or with size of village.

Employment opportunities

The people of Southwest Bougainville find work as labourers on plantations in Bougainville or elsewhere, work either locally or in more distant areas for the Administration, missions, or private enterprise, or for BCP, the major employer of labour in Bougainville.

Men from Southwest Bougainville have been attracted to the copra and cocoa plantations of the southeast coast of Bougainville or to those in the Gazelle Peninsula with which the

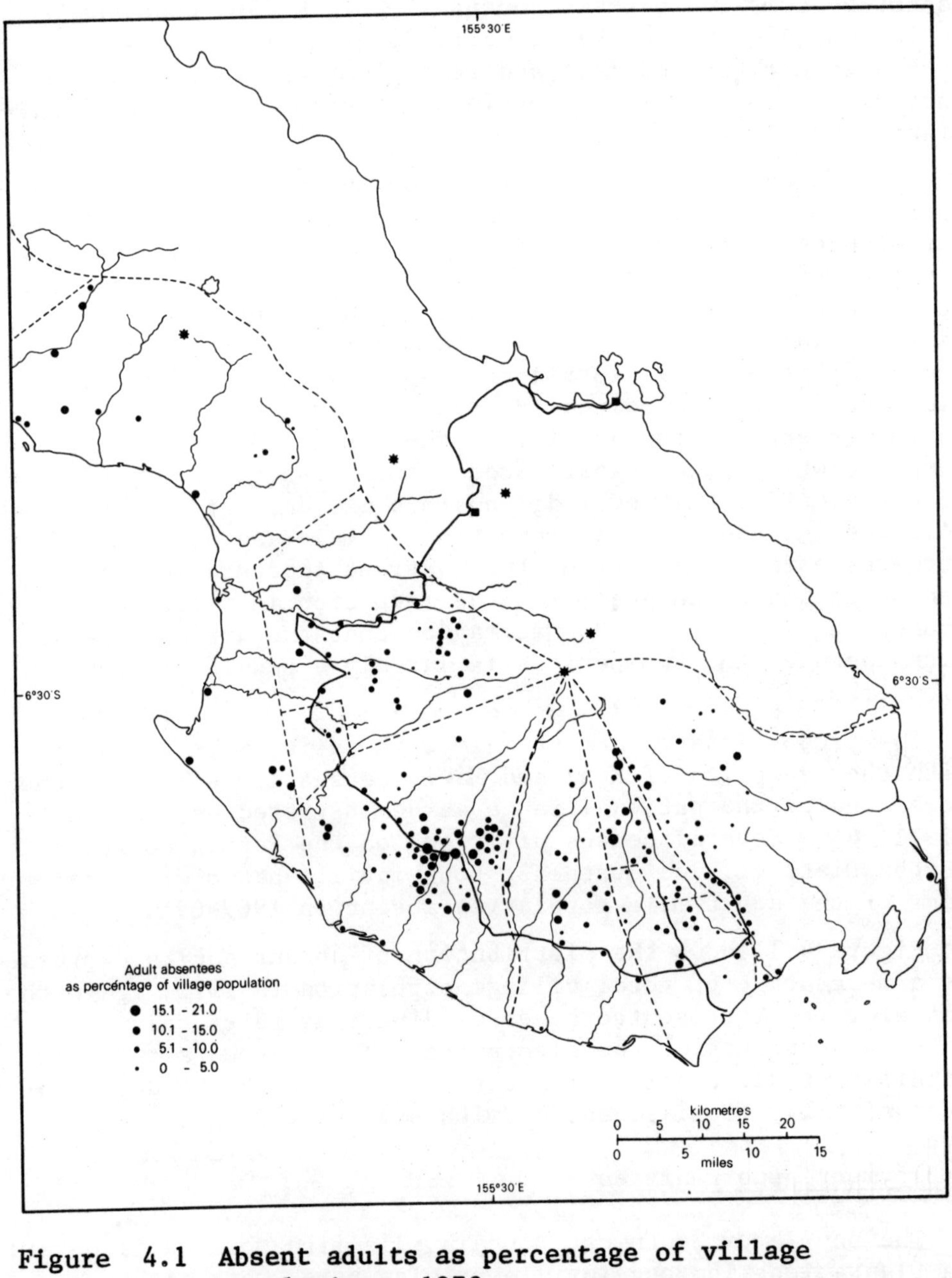

Figure 4.1 Absent adults as percentage of village populations, 1972

trading links and hence local shipping services were until recently strong. This employment opportunity was more important in the past, and at present no more than one or two hundred men at most are engaged in this form of outside wage-earning.

Work in government, for the missions and for private employers apart from BCP occupies more of the absentees than does plantation labour at the present time. These forms of employment offer scope for a much wider range of skills and educational achievements than does plantation labour. Because mission education was relatively well developed in Bougainville compared with some other areas of the country, it appears that there is a relatively larger number of Bougainvilleans with higher levels of skills and educational attainments than there are people from some other areas; they are in demand throughout the country. The following list of occupations of people absent from Siwai in 1970-71 gives some indication of the wide variety of jobs which absentees are filling: teacher, nurse, clerk, agricultural assistant, co-operatives officer, welfare assistant, radio announcer, medical assistant, surveyor, maternity and child health nurse, aidpost orderly, Pacific Islands Regiment, Royal Papuan and New Guinea Constabulary, catechist, nun, Marist brother, minister, pastor, driver, mechanic, sawmill employee, electrician, plant operator, mess assistant, stevedore, storekeeper, carpenter, painter, cook, casual labourer (Konga Patrol Report, No.1 of 1970-71).

The generally wide availability of primary education[1] has enabled a fairly large proportion of students to continue into secondary and in some cases tertiary education. The 1972 total for Bougainville District was 1,299 in secondary schools and 330 in vocational centres (Central Planning Office, Districts Supplement, 1973:172). In 1970-71, 206 students from Siwai census division were receiving post-primary education. The institutions they were attending are given in Table 4.3.

These comments should not be taken to imply that there is no need for further expansion of educational opportunities. There undoubtedly is a need for more, particularly in secondary and vocational training.[2] When further educational

---

[1] In 1972 Bougainville was one of only four districts which had an estimated 100 per cent of primary school places in relation to the number of children aged 7-12 (Central Planning Office, 1973:37).

[2] It is proposed to locate a new high school in the Siwai area.

Table 4.3

Numbers of students from Siwai census division attending post-primary institutions, 1970-71

Buin High School	42
Buin Vocational School	22
Hutjena High School	28
Kihili Girls' Vocational School	23
Other high schools	42
Popondetta Agricultural Training Institute	4
Teachers' Colleges	7
Other vocational schools	14
UPNG	8
Papuan Medical College	1
Seminaries, all denominations	7
Institute of Technology, Lae	8
Other institutions	5
Total	211

Source: Konga Patrol Report No.1 of 1970-71.

institutions are established in Southwest Bougainville they should be located at such centres as Beretemba, Amio or Konga.

The third source of employment for people from Southwest Bougainville is BCP, mainly at the mine site at Panguna but also in associated works at Arawa. In October 1972 some 413 persons, about 20 per cent of all absent adult males from Southwest Bougainville were employed by BCP. Table 4.4 lists the numbers from areas within Southwest Bougainville, and Fig.4.2 locates the home villages of these workers.

Table 4.4

BCP labour from Southwest Bougainville, 1972

Census division	Number of workers	Percentage of adult male population
Banoni	26	4.2
Nagovisi	91	5.2
Siwai	138	6.6
Makis, Lugakei, Paubakei and Kono	158	4.5
Total	413	5.2

Source: BCP labour records.

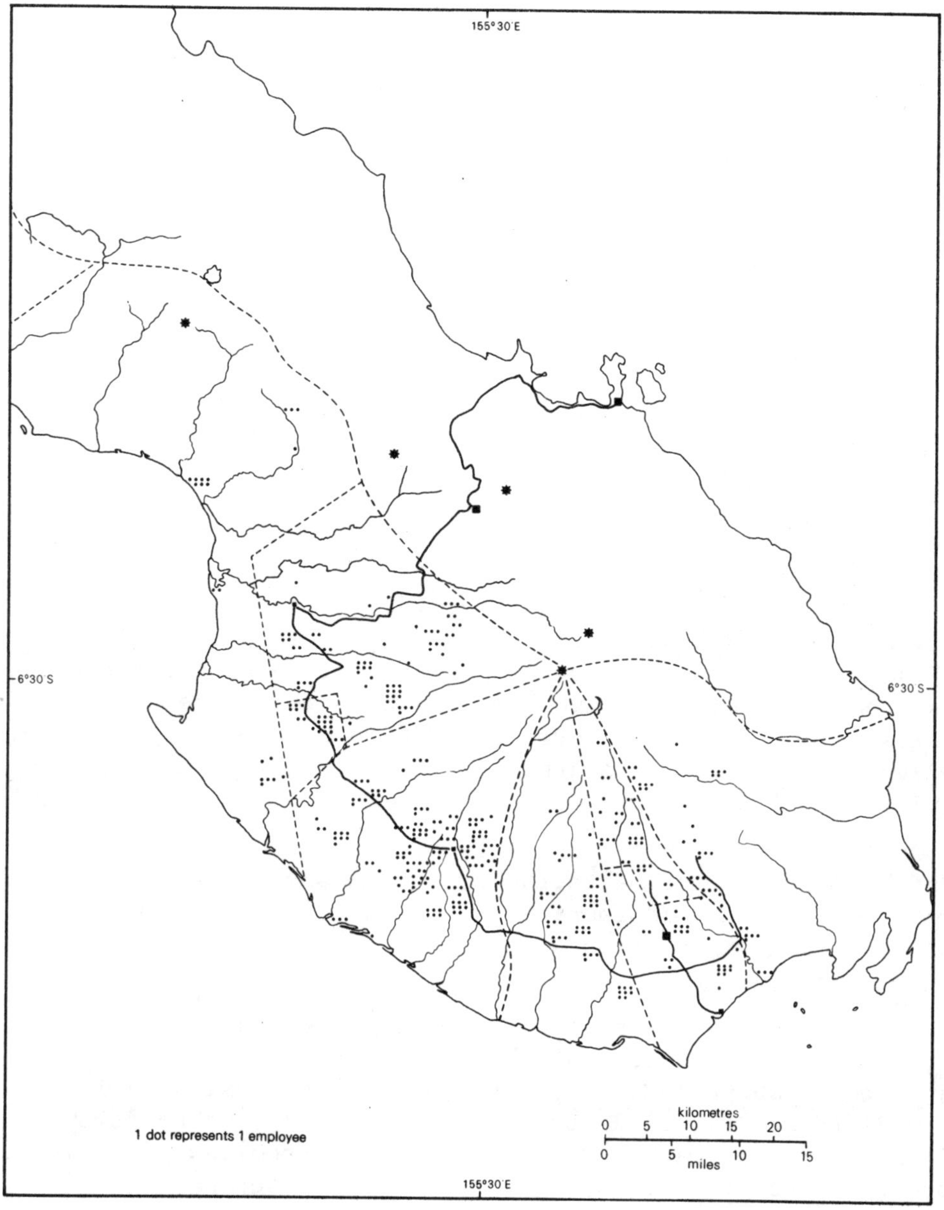

Figure 4.2 Numbers of BCP employees by village, 1972

People from Southwest Bougainville comprised about 39 per cent of the total Bougainvillean labour force of BCP in 1972, and 14.5 per cent of the total indigenous labour force of 2,850 (BCP labour records). They were drawn very widely from all parts of the Southwest, with no marked predominance of areas near Panguna. Some 162 of the 230 villages (70 per cent) were represented, by numbers ranging from 1 to 9 (from the large village of Tugiogu, Makis census division, which had 28 adults of its total population of 311 absent at that time).

Income contribution from absentee labour

Little information exists on the amount of money returned to the villages by absentees. However, one estimate of an average of about $300 per worker per annum has been made for Siwai (Konga Patrol Report No.1 of 1970-71). This would imply savings of about $6 per week, which would be very high for people on the weekly basic wage of $10-12. As noted above many absentees from Southwest Bougainville are at least semi-skilled, and about one-sixth of the absentees are working for BCP at Panguna where the opportunities for earning rates well above basic levels are high. However, in the subsequent tables a lower average figure of $100 per annum is taken as an average for income returned to the villages by absentee workers.

On this basis Table 4.5 has been constructed to show estimated annual income to the area from absent labour. If these amounts are incorporated with those derived from agricultural and compensation sources the figures given in Table 4.6 are obtained for total regional and per capita incomes.

This table indicates an average total per capita income for the region of about $32 which is lower than an independent estimate of about $48.50 per capita for the Siwai area in 1970-71 (Konga Patrol Report No.1 of 1970-71).[1] The figures in Table 4.6 indicate that the Siwai area receives slightly more than the Bana area per capita, and the southern third slightly less than the Bana area per capita, while the Bana area approximates the average.

[1] This estimate allowed $300 per worker for the returns from absent labour.

Table 4.5

Estimated income from absent labour, by census division

Census division	Income returned at $100/absentee p.a.
	$
Banoni	21,800
Nagovisi	46,600
Baitsi	7,200
Siwai	77,800
Makis*	24,600
Lugakei*	18,700
Paubakei*	12,000
Kono*	25,000
Southwest Bougainville	233,700

* Only those 'outside Electorate'.

Table 4.6

Incomes from agriculture, compensation and absent labour, c. 1972

L.G.C. area	Incomes from agriculture	Compensation	Absent labour	Total	Total/ capita
	$	$	$	$	$
Bana	148,596	100,000	75,600	324,196	31.75
Siwai	213,004	-	77,800	290,804	33.98
Buin	287,720	-	80,300	368,020	29.92
Southwest Bougainville	649,320	100,000	233,700	983,020	31.64

Source: Table 3.24 and pp.86-7.

Prospects for labour shortage in cocoa growing industry

It has been demonstrated in Chapter 3 that shortage of land is unlikely to limit the expansion of cocoa growing for several decades. However, many people think that the expansion of the industry will be limited by a shortage of labour in the near future. This question merits examination, but information which would allow detailed consideration is not readily available.

The question of whether there will be a labour shortage in the cocoa industry hinges very largely on the number of trees which can be managed by a grower and his family. DASF extension manuals suggest that 4 hectares of Class II land are needed to support a family by cocoa production. If 0.8 hectare is allowed for house and garden plots the remainder carries 2,000 cocoa trees at recommended spacing of 620 trees/hectare. This implies that a family unit can manage 2,000 cocoa trees, but this figure is far above the numbers yet planted in Southwest Bougainville. At 1972 prices to the grower 2,000 trees would earn a family unit $600 per annum or $92 per capita, assuming current yields of 2.7 kg of wet bean per tree.

Estimates of current plantings per average resident family unit (of 6.5 persons) are shown in Table 4.7. It should be noted that a considerable proportion of trees are still immature.

At first sight it appears that there is not a current labour shortage as the number of trees per average resident family unit is well below the DASF figure of 2,000 trees per holding. If it were assumed, arbitrarily, that 80 per cent of the population were directly dependent on commercial cocoa farming, the present village population of Southwest Bougainville might maintain some 8,000,000 trees. This represents a three-fold increase over present plantings and suggests there is no immediate labour shortage. However, this calculation is based on the hidden assumptions of effective mobility of labour within the area and equal access to land. Neither of these assumptions is justifiable and there is no doubt that serious (if localised) labour shortages would appear long before average plantings approached 2,000 trees per family unit.

There is some evidence that some people in Southwest Bougainville are already experiencing difficulty in maintaining and harvesting existing plantings (T. Moulik, pers. comm., January 1974). Whether this reflects an absolute shortage of labour or an assessment by cocoa growers that increased expenditure of time is not worthwhile compared with the alternative of allocating time to leisure or other pursuits is not clear, and would be an important topic for further research.

If the average family unit cannot maintain 2,000 trees and if for the sake of argument 1,000 trees represents a more realistic or acceptable workload, then the question of whether $300 represents a satisfactory cash income per family *vis-à-vis* cash incomes which can be earned through wage labour elsewhere becomes an important issue.

At present the highest average holding in the three subareas is 903 trees per average family unit in Bana. Undoubtedly

Table 4.7

Number of cocoa trees managed per the average resident family unit, 1972, by co-operative society areas

	Bana	Siwai	Buin	Southwest Bougainville
Total population, 1972	10,210	8,559	12,301	31,070
No. of all absentees, 1972	2,432	2,269	899*	5,600
Thus 'resident' population, 1972	7,778	6,290	11,402**	25,470
No. of average resident family units (a.r.f.u.) at 6.5 persons each, 1972	1,196	968	1,754	3,918
Approx. no. of cocoa trees, 1972	1,080,000	500,000	1,300,000	2,880,000
Percentage mature in 1972 (i.e. mature trees + new plantings to 1967)	28.3	74.8	32.8	38.1
No. of trees managed/a.r.f.u., 1972	903	516	741	735

* Only those 'resident outside electorate'.
** Numbers resident within electorate.

this average masks great variation between growers, and the reported labour shortages may relate only to those growers with larger plantings. If so, the chief labour problem may be that of transferring locally 'under-used' labour to those holdings where it is needed. In a semi-commercial economy where labour is not usually mobilised on a wage basis within the community, this may be difficult to achieve. The problem is aggravated by the fact that the main labour shortage occurs during the 'flush' when all growers are involved in harvesting their own cocoa.

The current planting rates for cocoa appear to be about 6.6 per cent per annum in Siwai, and much higher in the Bana and Buin areas. If planting continues even at a rate of about 7 per cent per annum it appears reasonable to believe that many more planters will experience a labour shortage in the not too distant future. This is likely to be most marked in the Bana area, then in Buin, and lastly in Siwai.

A number of possible solutions can be considered: the first would be to reduce present rates of absent workers. This would have little effect since the number of absent adults is relatively small and the contribution of their earnings to the community is probably equal to or greater than the monetary contribution they could make by producing cocoa. The second would be to employ wage labour from outside the district. In view of the existing antipathy of the people of Southwest Bougainville to large numbers of strangers in their midst this seems unlikely to happen although the presence of more adult workers, perhaps on short-term contract, would help ameliorate a labour shortage.[1] The third solution is to increase the productivity of the existing labour force. It is well known that much cocoa remains unharvested in areas of poor access (see Fig.3.4) because a high proportion of the available labour must be devoted to carrying the beans to the road. If roads were built into these areas, and collection services provided closer to the farms, then labour might be transferred to picking and hulling with a resulting increase in production in these areas. It is possible that improved agricultural techniques may help increase yields, and if so, spreading information and technical assistance should be a prime activity of the agricultural extension service.

---

[1] Since the antipathy of Bougainvilleans appears to be largely directed towards Highlanders, whom Bougainvilleans see as aggressive, a solution might be found if seasonally migrant labour were sought in lowland or island areas such as the Trobriands where a labour surplus exists.

The question of motivation is important but as yet little is known about it in Southwest Bougainville.[1] The desire to increase cash income may not necessarily be dominant and alternative uses of time and labour may be considered preferable to increasing cocoa production. On the other hand some informants stated that the desire to provide security and an inheritance for their children was a reason for expanding their cocoa holdings.

The opening of the trans-island road in Southwest Bougainville will allow increased ease of movement for absent workers. People employed at Panguna, Arawa and Kieta or the plantations of the southeast coast will find it possible to visit their home villages more frequently since the journey can be made by passenger motor vehicle (PMV) to Nagovisi and Siwai in about half a day and to Buin in a day. The cost will be much lower than the air fare. This fact may encourage a higher proportion of younger men to seek work in central and southeast Bougainville particularly in Panguna, Arawa and Kieta. On the other hand many absentees presumably already own cocoa plots in their villages, and the road may encourage them to return more often than before to maintain their young cocoa and harvest the crops. The road is likely to increase the mobility of labour especially from Nagovisi and Siwai, but whether the result will be an increase or reduction in the labour available for cocoa production is uncertain.

The presence of the road also introduces new earning or business opportunities, through the increased demand for PMV services, service stations, mechanical repair work, and for food, drink and accommodation services which it wil induce. Thus it will introduce a new element of diversification into employment and earning opportunities within Southwest Bougainville.

---

1 Current studies by Dr T. Moulik and others should provide valuable information on this topic.

Chapter 5

Urban development

Southwest Bougainville is essentially a rural area, the only place designated as a town being Buin in the far south. Until the development of Panguna and Arawa after 1968 the only place in southern Bougainville which could properly be called a town was Kieta which had earlier been the administrative and port centre of the island. After World War II the administrative headquarters was transferred to Sohano on Buka Straits in the north, which was also a port of some size. Kieta did not again become the administrative headquarters for Bougainville District until the mining developments of the late 1960s changed the focus of economic and political activities to central Bougainville. In the last few years there has been very rapid urban population growth at Panguna, Arawa and Kieta. Table 5.1 lists 1966 and 1971 populations for these three urban centres and Buin.

By 1971 Buin had 600 people, while Kieta had 2,400. Kieta, Arawa and Panguna together had an urban population of some 14,000 in 1971, and this figure was expected to rise to about 16,500 in 1971 then fall sharply to about 10,000 in 1972, and thereafter to show a steady but less rapid increase to about 17,000 by 1980 (DASF, 1971:9 and Appendix p.iv).

Growth rates were much higher in Kieta, Arawa and Panguna than in Buin between 1966 and 1971 as Table 5.2 shows. The growth rate of Buin is only slightly above the national average urban growth rate of about 9 per cent per annum between 1966 and 1971[1] (Ward, 1973:271) while Kieta was one of the fastest growing towns in the country, as were Arawa and Panguna which are entirely new towns. In Buin the growth rate of the indigenous section of the population was much higher than that of the non-indigenous section, while the reverse was true for Kieta, and probably also for Arawa and Panguna.

---

[1] Calculated on preliminary census figures.

Table 5.1

Urban populations, 1966 and 1971

Year	Urban centre	Indigenous population			Non-indigenous population			Total		
		M	F	Total	M	F	Total	M	F	Total
1966	Buin	207	99	306	22	17	39	229	116	345
	Kieta	478	166	644	79	32	111	557	198	755
1971	Buin	392	171	563	30	20	50	422	191	613
	Kieta	1,462	332	1,794	433	175	608	1,895	507	2,402
	Arawa*	2,722	404	3,126	1,528	388	1,916	4,250	792	5,042
	Panguna**	3,583	87	3,670	2,406	506	2,912	5,989	593	6,582
1971	Urban total: Kieta, Arawa, Panguna	8,159	994	9,153	4,367	1,069	5,436	12,134	1,892	14,026

* Includes Loloho
** Includes Wairovi.

Source: Bureau of Statistics, 1972: Urban Centres, Konedobu (preliminary figures).

Table 5.2

Urban growth rates, 1966-71

	Indigenous	Non-indigenous	Total
	% p.a.	% p.a.	% p.a.
Buin	12.97	5.09	12.18
Kieta	22.74	40.51	26.04

Source: Calculated from Bureau of Statistics, 1972.

Buin was in 1966 and even more in 1971 the most 'indigenous' of the four urban areas, as Table 5.3 shows. As with most towns in Papua New Guinea, all the south Bougainville towns had high ratios of males to females in their population, but Buin was the least aberrant in terms of this measure (see Table 5.4).

Table 5.3

Indigenous proportion of urban populations, 1966 and 1971

	1966 per cent indigenous	1971 per cent indigenous
Buin	88.7	91.8
Kieta	85.3	74.7
Arawa	-	62.0
Panguna	-	55.8

Source: Calculated from Bureau of Statistics, 1972.

Table 5.4

Ratio of males/100 females in urban populations, 1966 and 1971

		Indigenous	Non-indigenous	Total
1966	Buin	209	129	197
	Kieta	288	247	281
1971	Buin	229	150	221
	Kieta	440	247	374
	Arawa	674	394	537
	Panguna	4,118	475	1,010

Source: Calculated from Bureau of Statistics, 1972.

The imbalance is more strongly marked in the indigenous section of the population in all towns, and extremely so in Panguna which at the time of the census was almost at the height of the construction phase. The sex ratio in the towns can be expected to tend towards a more normal even ratio as more women and children join the men in the towns, but the process will be relatively slow, and imbalance is likely to remain for many years.

Form and function of Buin town

Like most other towns in Papua New Guinea Buin originated as an administrative settlement. Unlike some other towns the airfield does not form the focus of the settlement, although it is only three miles away. The nucleus of Buin town is a small grid street pattern built on level land with pre-existing villages nearby but not incorporated in the town (Fig.5.1). The District Office occupies a central position, with associated high-cost European style government housing near it. Close by also are the offices of the co-operative societies, with the European club and indigenous club buildings within short walking distance. Commercial premises are located on the 'main' street and around one block adjacent to it. Further west along the main street and away from the District Office are the police and administration labour quarters, a small hospital, the PWD depot and government transport pool, and further south the DASF station. At the eastern end of the town are Buin High School and the town market, while a mile or two further east is the co-operative society's cocoa fermentary.

No breakdown of employment figures exists to show the functions of Buin town, but these are by now somewhat more diversified than at the time of its establishment. Administration is still dominant since the town is the headquarters for Buin subdistrict. As such it houses officials and staff of a number of government departments, which include District Administration, Police, Public Health, Education, Public Works, DASF, and Business Development (mainly co-operatives division). It is also the headquarters of Buin Local Government Council. Buin town is the main commercial centre for southern Bougainville, and this function is represented by the processing and wholesaling activities of the co-operative societies, and their organisation of the stevedoring required at Kangu port, and retailing and trading by European and Asian entrepreneurs. As the base of the one construction firm operating in Southwest Bougainville it has a small building and construction function. Through the presence of the airfield and agencies of the airline

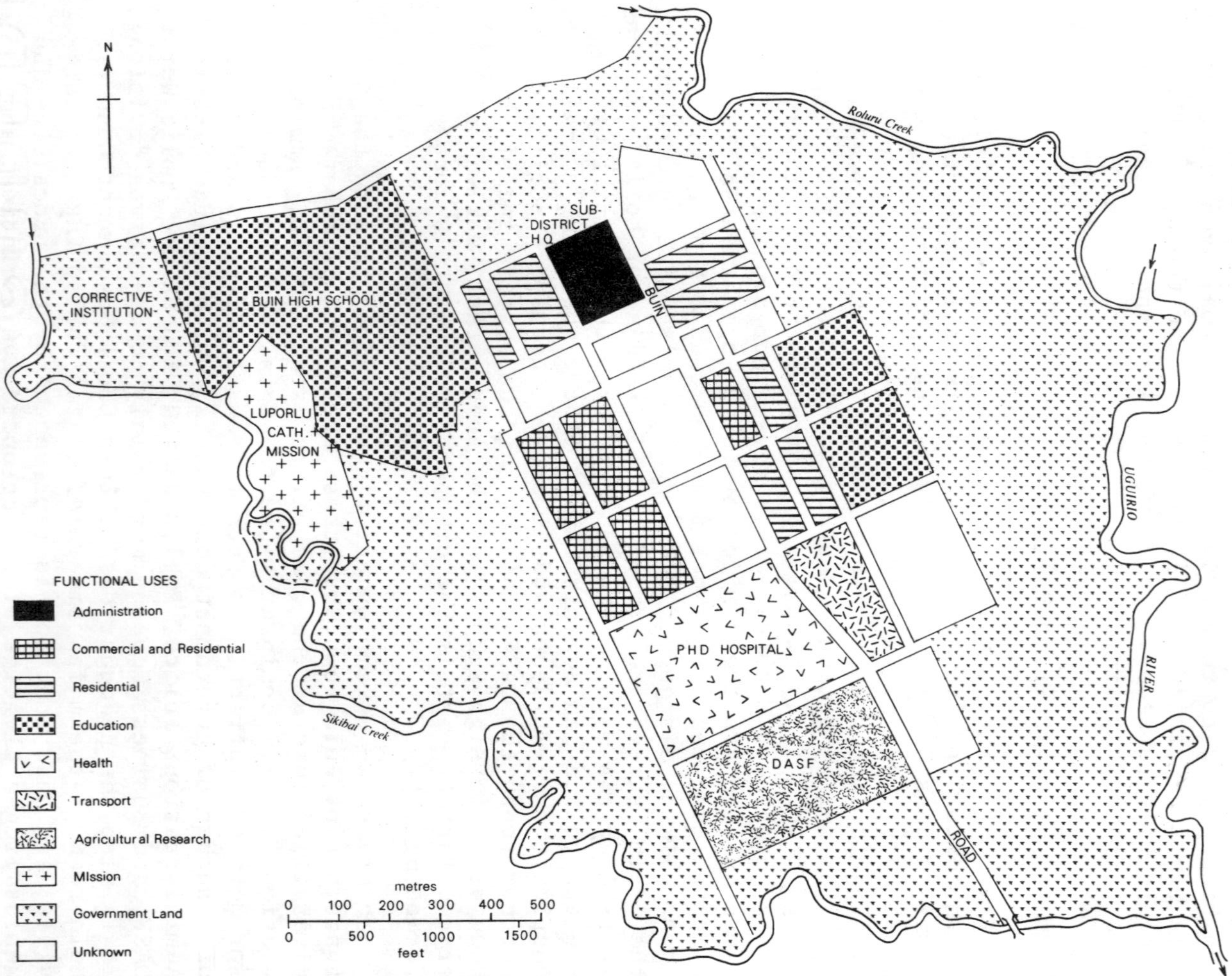

Figure 5.1 Land use in Buin Town, 1972

companies and their associated small accommodation facilities, and the presence of the government road transport servicing depot and transport pool it has a transport and service function; most PMV activities however are located not in Buin town but in villages throughout the area. The educational function is important, with a secondary school and at least one vocational school located in the town. There is a hospital at Buin, though the mission-operated hospital at Moratona in Nagovisi is much larger, and there are other small mission hospitals in the subdistrict.

Most of the numerically small European population of the town are employed in the various government services and as school teachers; a few are in commerce and related activities. The government is also the main employer of indigenous persons in the town, the majority of the employees occupying semi-skilled or unskilled positions such as clerks, drivers, police, labourers, and DASF assistants. Most of the female indigenous employees are teachers, nurses, typists, or shop assistants.

In short, Buin in 1972 had many of the attributes of a typical small outstation centre in Papua New Guinea, established where transport facilities were relatively good, but it was not entirely central to the population of the area it was intended to administer. It had marked social divisions between the indigenous and expatriate sections of the population, which were heightened by the differences in living styles and incomes of the two groups.

Incipient urban centres

Although no other places in Southwest Bougainville are formally recognised as towns there are a few locations which may well develop into towns in future. The most notable of these are Beretemba in Nagovisi, Konga and the site of the Siwai Local Government Council chambers near Amio in Siwai. In the south such possible future centres are less obvious, but possibly Piano in Makis and Tabago in Kono may develop further in future. Both are important mission centres while Piano also has an exceptionally large trade store and is the home of an important indigenous leader.

Beretemba is located on the main road fairly close to the centre of population of Nagovisi and Baitsi census divisions. It is now the site of the Bana Local Government Council chambers and of the main cocoa fermentary of the Bana Producers' Co-operative and the main store of the wholesaling co-operative. It is an obvious location for future development of indigenous transport-related industries, such as service stations and motor vehicle repair works, restaurants, cafes, guest houses, and trucking companies. As government services expand in Southwest Bougainville, Beretemba could be the site of a high school, a vocational training centre, community development facilities, agricultural extension offices, business development offices and so on. The main hospital for the northern area will probably remain at Moratona mission as it is readily accessible, about six miles northwest of Beretemba by good road.

In Siwai two places share potential for further urban development. Konga has in the past been an administrative sub-centre, and at present has the major agricultural station for the Southwest and is the base of a rural development officer. It also has a rural health clinic, a large and active primary school which could form a nucleus for further community development, and is the headquarters of the Siwai co-operatives. About three miles away near Amio village are the Siwai Local Government chambers. There must have been good reasons for the choice of this location for this purpose (possibly the location is slightly more central to the Siwai population) but Amio is off the main road, although close to one of the main mission centres in Siwai. Amio is thus less favourably situated than Konga for the development of transport-related industries, and seems slightly less likely to develop than Konga, provided adequate land is available at Konga.

In 1972 the only third-level government station in Southwest Bougainville was Boku Patrol Post, located on a river terrace bank high above the Puriata River on its southern side. While the location is roughly equidistant between the relatively dense Nagovisi and Siwai populations it is in fact rather remote from both. There has been no development of ancillary functions such as retail trade at Boku Patrol Post. In view of the fact that the site is likely ultimately to be eroded by the Puriata River it would seem reasonable in future to abandon Boku Patrol Post and divide its functions between say Beretemba and Konga or Amio. The Boku airstrip, which is located on the north bank of the Puriata, could still be readily reached by road from Beretemba, particularly in flood emergencies. The importance of the airstrip is anyway likely to decline with the opening of the new road, except for emergency use.

In the south, Buin town is likely to remain the major urban centre. It is relatively central to the populations of Makis, Lugakei, Paubakei and Kono census divisions, and it is likely to retain its role as a focus for roads to the west, north, east and south, and to retain its airport for some time to come. The development of any other lower level centre in the south is likely to be restricted by the existing dominance of Buin town. Buin town could be the base for mobile service units circulating in the southern third of Southwest Bougainville (see R.G. Ward *et al*., 1974).

Effects of the new road on urban development

Because of their favourable location, Beretemba and Konga and to a lesser degree Amio stand to benefit considerably from the opening of the trans-island road to serve the increasing amount of traffic which will pass over it. The orientation of the Southwest Bougainville economy is likely to change from its former southern direction to Buin and Kangu to a northern and northeastern direction to Kieta. This will apply particularly to Nagovisi and Siwai. Hence Buin town will experience a reduction in its relative importance as roughly two-thirds of the movements of exports and imports which formerly passed through Kangu are diverted to the north. Since Buin already possesses a limited range of service facilities the immediate expansion of service industries is more likely to occur in Beretemba and Konga than in Buin. The northern two-thirds of the main road will in future carry probably more than half the total traffic in contrast to the former position when the southern third probably carried more than half the traffic volume.

The change in orientation of trade is also likely to have effects on the urban centres outside Southwest Bougainville. There is likely to be an increased amount of short-term movement of people from Southwest Bougainville to Panguna, Arawa and Kieta. Panguna is likely to receive an increased flow of sellers of fresh fruit and vegetables from Nagovisi. While this will not affect the size of its permanent population it is likely to result in increased sales at the supermarket and trade stores located there.

Three of the co-operative societies in Southwest Bougainville are investing in relatively large shop developments in Arawa town centre, made possible in part by Development Bank loans. These developments are partly related to the construction of the trans-island road, and now that it is open it will encourage people from the Southwest to visit Arawa to inspect and purchase from these stores. If this type of movement occurs it may well contribute to the decline of small local trade stores in the Southwest.

Kieta stands to benefit because virtually all of the cocoa and copra produced in the Southwest will now be shipped overseas from that port, in contrast to the pre-road situation in which at least some produce was shipped from Kangu to Rabaul for export. This will mean some increase in the ancillary services associated with cocoa shipment, such as road transport activity, storage facilities, marketing and cargo handling. Until better vehicle sales and servicing facilities are established in the Southwest, businesses which supply these services in Kieta and Arawa also stand to increase their volume of activity.

In general the main consequences of the opening of the trans-island road on urban development will derive from the reversal of the orientation of the economy towards the north, and the increased mobility which the road will offer. The latter will be linked with increasing investment in motor vehicle ownership, which can be expected to show a sharp rise in the first one or two years, probably followed by a levelling-off period. Without supplementary investment in service facilities, driver and mechanical education, and business economics of running PMVs or truck transport, the disillusionment experienced in other areas where roads have been built or upgraded may well occur in Southwest Bougainville. The existing or incipient towns of the Southwest are the logical places to establish these services.

Chapter 6

Existing transport systems: roads and vehicles

The Buin and Siwai plains provide relatively easy terrain for building tracks and roads. In pre-European times it is likely that many villages were linked by footpaths. Trans-island foot trails almost certainly existed in the northern Banoni area where ethnic groups overlap from the east to the western side of the main ranges.

These tracks formed the basis for the 'roads' which the German and pre-war Australian administrations required the people to maintain between villages. The line of the major road of the Southwest, linking Kangu to Moratona, was laid down in German times, and has remained in use to the present, although in Siwai census division it was replaced as the main road by a new road completed in 1970 about six km inland. Oliver (1955:19) describes the intervillage paths as they were in Siwai about 1938 thus:

> Pathways vary greatly in width: from faint trails through regions of secondary growth to eight-foot-wide roads - highways, actually - along which the Administration officials travel during their periodic census and tax-collecting expeditions. But width is no indication of the amount of traffic - the beautifully repaired 'Government Road' might remain untravelled for days on end, whereas the tiny track leading from village to spring will be in constant use. Real transformation of the landscape by roads has occurred only in those places where rival leaders have vied with one another to see who could clear the widest roadway.

Oliver refers also to the fact that these 'roads' were supposed to be 'straight trails some nine feet wide and suitable wherever possible for bicycle travel' (Oliver, 1955:324). The use of bicycles about 1927 by Roman Catholic catechists is noted by Laracy (1969:128).[1]

[1] Southwest Bougainville is unusual in Papua New Guinea in that bicycles are widely used for passenger travel as is shown

Bridges were ordered to be built across smaller streams, and these were required to be roofed over to prolong the life of the spans (Oliver, 1955:324). At least one such bridge with a well-maintained thatch roof remains in use today, crossing a tributary of the Hari River on the Amio-Mamagota road in Siwai.

Wartime roads

This pre-war 'road' and track network was utilised and improved by both sides during World War II (see Fig.6.1). The Japanese are known to have constructed Kara (now Buin) airfield; built a light railway to connect it with the port at Kangu and the former Australian administration post at nearby Kahili; and to have used Tonolei Harbour as a deepwater naval base. They probably also improved the condition of the roads in South Buin. The Americans, after establishing a base at Torokina, created a new dense network of vehicular roads connecting the four airstrips they built there, and serving the base. Most of this construction was allowed to decay after the war.

When the Australians took over Torokina and began to fight their way southwards into Buin they found

> there were native trails along the coastal plain some of which had been improved sufficiently to carry light vehicles. From Marawaka [Mawareka] southward, for example, the track had been widened to about twelve feet. In the interior a network of footpaths crossed the island over saddles in the mountain chain. One was the Numa Numa trail...connecting Torokina and the east coast, and the other connected the Jaba River with Kieta. (Long, 1963:104.)

---

1 (continued)
by the fact that in 1970 there were 439 bicycles in Siwai, or one bicycle to every 16 people (Konga Patrol Report No.1 of 1970-71, Incidental Statistics). There appears to be little or no use of bicycles for moving goods, though a simple basket attachment would readily provide a means. In contrast to motor vehicles, bicycles offer many advantages in that the area in general is relatively flat, the capital cost and running costs are low, maintenance is technically simple, bicycles can be easily lifted across fords, and they are much less dangerous than motor vehicles.

Plate 7. Covered bridge on Konga-Mamagota road

Plate 8. A fallen tree blocks the old Army and German road on the coastal plain, east of Runai, Siwai census division

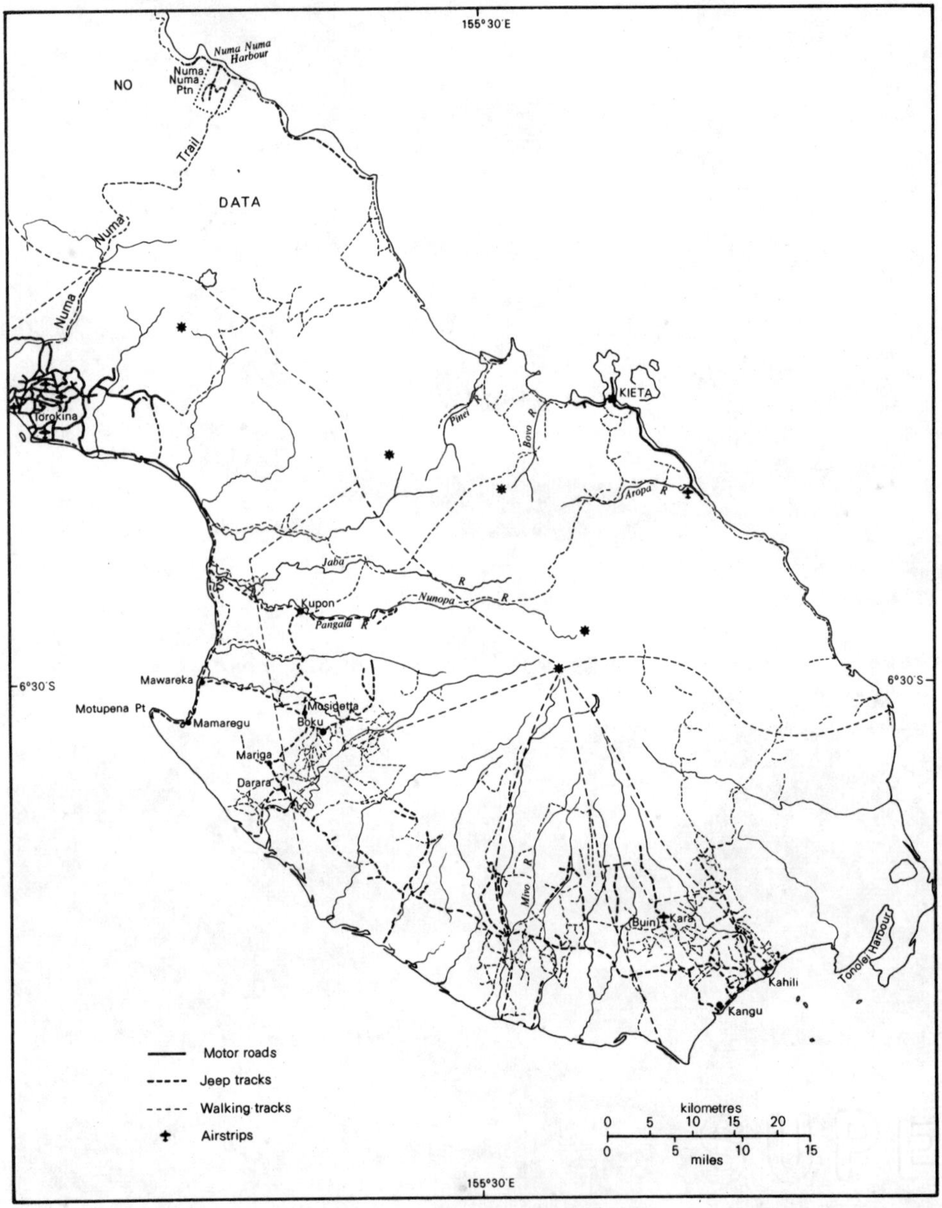

Figure 6.1 Roads and tracks, c.1944

There is no evidence of the latter track on the wartime map (No.3313, Bougainville Island South 4 Mile Series, 1945), but a jeep track is indicated following the Pangala and Nunopa rivers well into eastern Nagovisi, and it is probable that walking tracks did exist, following either the Nunopa headwaters into Kongara census division and thence down the Aropa valley to the east coast, or further north following the Jaba River headwaters into Guava census division and down either the Bovo or Pinei river valleys to the east coast.

The Australian forces found it 'necessary to make roads forward from Torokina to carry either jeeps or 3-ton trucks and in some cases both. Each such road involved laying extensive corduroy and building many bridges across rivers and creeks' (Long, 1963:177). The road south from Torokina followed close to the coast along the beach ridges of the Jaba land system, and a branch turned inland at the Jaba River which it crossed by a 'rope bridge' some four miles from its mouth. This route then followed up the Pangala River to Kupon, while another continued to follow the coast through Mawareka to Motupena Point.

From Kupon the Australians found

> an old 'Government road' led through the bush to Mosigetta. Like most such tracks in this area it was from 10-15 feet wide and bounded on each side by a deep and narrow ditch. [The ditch was still recognisable in October 1972 at a point further south (east of Kotu village) on a section of the 'government' and army road that has been very little touched since the war.] In the timbered country the trees met above the track forming a green tunnel, and the secondary growth was so dense as to hide a man standing a yard or two off the track. When the infantry began patrolling forward the tracks were still carpeted with grass, but the jeeps and trucks that followed tore this up and churned the surface into deep mud. At length the worst sections of each track had to be 'corduroyed' - paved with logs laid edge to edge - and over these the vehicles would lumber until, as the days passed, they pressed the logs deep into the mud. (Long, 1963:142.)

The Australians worked southwards in the later months of 1944 to August 1945 when the remaining Japanese forces surrendered to them on the banks of the Mivo River, about three-quarters of the way from Torokina to (present-day) Buin. They rebuilt the old 'Government Road', which ran 6-8 km inland from the coast, and constructed a new one, known as the 'Commando Road', roughly parallel and some 6 km further inland. Each had to be corduroyed to take heavy traffic. When the war ended,

> the main road from Torokina to the Mivo, 75 miles long, had been built through virgin coastal scrub on volcanic sandy soil, and through swamp forest, often on clayey soil, cut by many rivers and lagoons. In addition the engineers and infantry had built 21 miles of lateral roads, 25 miles of corduroyed jeep and tank roads, and more than 50 miles of tracks that were temporarily made capable of carrying tanks or jeeps and then abandoned. In the southern sector the engineers built 30 permanent or semi-permanent bridges totalling 9,500 feet. Frequent flooding made maintenance a heavy task, and the two major floods [in April and July 1945] caused great destruction and much earlier work had to be remade. (Long, 1963:225.)

The remarkably dense network of roads and tracks in South-west Bougainville about 1945 is shown in the extract from the wartime map in Fig.6.1. The basic elements of the present pattern can be traced, particularly the main connection from Buin to Kupon, and the numerous branches leading inland and to the main rivermouth villages on the coast.

Postwar deterioration

Without continual maintenance the Army roads had a very short life under the climatic conditions of heavy rainfall and frequent flooding. For example, a patrol report for the Mamaromino area (Paubakei census division near the present Buin town) reports that 'practically all roads and bridges [are] in a state of disrepair' (Bougainville Patrol Report No.3 of 1947 (Buin subdistrict)). The same report stated that very little work would be needed to make the Japanese airstrip at Kara serviceable, but that if it were reopened the ten miles of road connecting it to the government station (at Kahili) would need attention. This road had a Japanese-constructed narrow gauge 'tram line' along its full length at that time. The airfield and road were later reconstructed and the administrative post moved inland to near the airfield when Buin town was founded about 1962.

Similarly in February 1947 slightly further west in the Aku area of present Makis census division 'all except the main road to Siwai need attention even to make them fair pedestrian roads' (Bougainville Patrol Report No.4 of 1947). The road running inland from Kara airstrip was in better condition (as the 1945 map verifies): 'From Kukumaru rest house [about 2 km north of the airstrip] to Pariro and on to $1\frac{1}{2}$ hours walk from Koniguru [i.e. about 6 km) the road is jeepable' (Bougainville Patrol Report No.5 of 1946-47 (referring to April 1947)).

Further east in present Kono census division the Japanese had constructed 'well-formed and drained' roads to take vehicular traffic from Mituai to probably about Tabago. From Tabago east to Turitai the vehicular road had deteriorated to only a bicycle track by August 1947, two bridges having collapsed. A graded track had been cut by the Japanese from Turitai to Ububakogi, but 'the natives had made a short cut' (Bougainville Patrol Report No.7 of 1947-48). These roads or tracks are the basis of the present road to Tabago (in October 1972 in bad condition after Mugua village), and walking tracks from there northeast to the Oria area of Kono census division.

Plate 9. The feeder road between Buin town and Tabago village in good condition near Luaguo, Paubakei census division

Foot-tracks now as then also connected Tabago to Pariro, though they are not shown on either the 1945 map or the current district map.

In February-March 1948 the main road from Buin to Torokina (known as 'No.1 Road') was in good condition and (apart from collapsed bridges) was suitable for vehicular traffic as far as Darara (i.e. Jarara, just north of the Puriata River across the border into present southern Banoni census division), but from there on to Mariga and Manaregu (on the coast east of Motupena Point) it was swampy and impassable (Bougainville Patrol Report No.10 of 1947-48). By April-May 1949 the old Army roads in the Banoni and Baitsi areas were reported to be breaking up, their corduroy tracks rotting. The section from Mawareka to Mosigetta was 'almost done' (presumably meaning impassable rather than 'nearly reconstructed') (Bouganville Patrol Report No.1 of 1949). Vehicular passage was impossible, as it was over roads in the Nagovisi area (Bougainville Patrol Report No.2 of 1949).

Reconstruction and new road building

The non-availability of some Bougainville Patrol Reports makes it impossible to trace in detail the history of road maintenance and reconstruction in Southwest Bougainville in the 1950s. Some minor work must have been done but generally one has the impression that this decade saw little activity in improving transport facilities or towards the economic development of indigenous people in Bougainville. The 'No.1 Road' must have remained the major line of communication (and its use was constrained by the fords across the major rivers) while bicycles and foot must have been the main modes of transport. Sea communication with the outside world was through the port of Kangu, and the area was served from Rabaul, not Kieta. Bougainville Island was then administered from Sohano, at Buka Passage in the north, and the Southwest must have been very isolated.

It is not until the mid-1960s that it becomes possible to trace developments further. By this time cocoa planting had begun in the southwest, and the attention of the Australian administration had become firmly focused on the production of export crops such as coffee in the highlands, and copra and cocoa in New Britain, New Ireland and Bougainville. CRA began exploration at Panguna in 1964, thereby drawing some attention to Bougainville, though major consequences such as the move of the administrative headquarters to Kieta did not occur until 1968. The first House of Assembly was elected in 1964, and the second, which contained some strongly supported Bougainvillean representation, in 1968. The Directorate of Transport, whose task was to co-ordinate administration transport policy and planning, was established in 1967.

The documentary evidence for the 1960s is concerned with major transport improvements rather than minor ones. It is assumed therefore that while modest essential maintenance probably continued to keep minor roads and intervillage foot-tracks open there was no great effort towards the improvement of feeder roads in Southwest Bougainville until very late in the decade. Interest was, however, beginning to be shown in various alternative schemes which might provide better access for the area as a whole.[1] These included the suggestion of a mountain road from Kieta to Boku (in November 1963), the construction of a good quality road from Buin to Boku (which was later to materialise), and the suggestion of a road around the southeast of the island to link Kieta with Buin via Aropa, Toimonapu and Tonolei Harbour.[2] In 1964 a feasibility study for a trans-island road was carried out by consulting engineers who concluded that the cost of an Aropa-Boku road (some 54 miles in length) would be $2,350,000, an average cost of $42,000 per mile plus bridging costs. By March 1966 design work had been commenced on a road from Buin to Toimonapu, but the collapse of the Tonolei Harbour timber project about 1970 also put an end to hopes of an early construction of this road.

In 1967 negotiations between the administration and CRA led to an agreement that the company would supervise construction of a road from Panguna to Moratona, and that within two years of the commencement of construction the administration would reimburse CRA to the maximum extent of $500,000 for this work. The standard of the road to be constructed was to be regulated by consultation with and inspections by CRA and the Department of Public Works. This agreement, however, was never ratified by the Department of Territories in Canberra, and apparently the company lost interest in this project as

---

[1] I have briefly traced the course of decision-making towards the solution of Southwest Bougainville's transport problems in Ward 1971:308-11.

[2] The information in this and succeeding paragraphs is mostly drawn from a Report on the History and validity of proposals for a road, Panguna to Lambalam/Beretemba, South Bougainville, held in the files of the Department of Transport, Port Moresby.

Plate 10. Bridge over the Puriata River at Boku Patrol Post, Baitsi census division, on the main road

Plate 11. The treacherous sandy-bottomed ford across the Niwo River in Makis census division. The line of the crossing runs directly across the centre of the photograph. Note flood debris

early as 1967, when it found that its needs could be met by a road to the new port at Anewa Bay on the east coast.

Meanwhile political disaffection in Bougainville generally but including the Southwest, was mounting, and finding expression in particular through one elected member, whose strong power base was located in the Southwest. In the Southwest smallholder cocoa production was increasing and demands for a better means of export were strong. To appease both these conditions funds were approved for the then highest priority item of the proposed major road developments in Southwest Bougainville, namely construction of the Boku-Konga road. This road, now referred to variously as 'Barclays' road' (after the construction firm concerned), or by local people as '_rot wara_' (a reference to the fact that, like a lake, it doesn't lead anywhere in particular), was constructed between late 1968 and August 1970, roughly along the line of the 'Commando Road', through the heart of the most densely populated part of Siwai census division. Its cost was $776,000. It includes one major bridge, over the Puriata River at Boku Patrol Post, but its inland route enables it to cross three other major rivers (the Hongorai, Hari and Ogorata), as well as a number of tributaries, higher up their courses with bridges of relatively moderate length and cost. The immediate effect was to provide improved access to Buin town and Kangu port for Nagovisi, Baitsi and Siwai census divisions.[1]

---

[1] Some of the incidental short-term benefits of the construction of this road are interesting and have been identified (in Konga Patrol Report No.1 of 1970-71) as follows:

Direct payments to local labour over 20-month construction period: $45,500.
Revenue to local people from food and incidental purchases by imported labour: $43,000.
Vegetables purchased for expatriate mess over construction period: $2,300,
i.e. an estimated cash injection into the local economy of $90,800.

In addition the construction company established a favourable impression with local people in that it paid a higher rate for casual labour than that offered by the Administration; collected and delivered local labour; trained local people; demonstrated the necessity to keep road-making machinery in operation by repairing breakdowns quickly; and willingly paid minor damages to local people for accidental destruction of property. (continued)

A feasibility study carried out in 1967, which preceded construction and decided the route for the Boku-Konga road, continued through Makis and Paubakei census divisions to Buin, again following a higher route some six km inland from the existing main road. This section, however, remains to be built. As further west, the inland route avoids the major river crossings (at present sometimes dangerous fords) on the Mobiai, Mivo, Oamai, Porror and Silibai rivers. Its construction would particularly benefit the southern third of Buin subdistrict, especially by removing the uncertainty and potential danger of the existing fords.

Meanwhile the trans-island route held low priority for road needs in Southwest Bougainville, the Konga-Buin, and Boku-Moratona sections being regarded as more important. In November 1967 the Panguna-Lambalam-Jaba road was dropped from the Administration's current Bougainville planning (Dept of Transport Files, see footnote 2, p.119). At this time the subdistrict headquarters at Buin pointed out the need for upgrading the Jaba-Lambalam road, and favoured the idea of shipping local produce through a port on the west coast - presumably Jaba, Mawareka or Mamaregu, all surf ports lacking in facilities or road access. Another suggestion was that the airstrip at Boku should be upgraded to DC3 standard for the shipment of produce (presumably fresh vegetables and cocoa were intended), but no action was taken towards either the upgrading of west and south coast ports or of the airfield.

The suggestion of the trans-island road was raised again early in 1968 when the then District Commissioner for Bougainville sought further information. A meeting between CRA and

---

1 (continued)
After the contract was completed an expatriate employee of the company remained in Southwest Bougainville with a bulldozer with which he worked improving local feeder roads for Local Government Councils for a time (thereby winning much goodwill), before finally 'walking' the bulldozer back to Kieta overland through Nagovisi and Panguna, creating the first trace of a trans-island road, and becoming the first person to drive a vehicle across Bougainville.

An indication of the extent to which local people were prepared to support feeder road improvement is given by Smith (1972:12) who reports that in 1970-71 the President of Buin Local Government Council personally collected $27,000 in cash in four days from the people for bulldozer hire, and the people added to this 24,000 man-days of voluntary labour and 3,000 hours of tractor hire.

the Assistant Administrator in February 1968 apparently discussed the possibility of extending the road from Panguna to the west coast using temporarily idle CRA equipment, provided the Administration would contribute $50,000 or $60,000 to the cost. Again the suggestion did not result in action.

In November 1967, however, the Directorate of Transport obtained an order of cost estimate for a road from the Jaba River Pumping Station to the west coast, which resulted in a figure of $750,000 for a road of Rural Class II standard. This relatively high figure was apparently caused by the need to bridge at least one major river (the Pangala), and because the road would need to be of moderately high standard since it would become part of a trunk road system. The Directorate of Transport benefit/cost analysis of this proposed link resulted in a ratio of 0.930 (benefits $896,681, costs $963,696), which, although less than 1, would not necessarily have precluded construction. At this time the Directorate of Transport had doubts about whether the standard of the Company's road between Panguna and the Jaba River Pump Station would be adequate for public use. The standards proposed for this section at that time were: average maximum grades 15 per cent; minimum radius curves 50 feet (15 m); average road width 12 feet (3.6 m). These were considered to be too dangerous for public use unless major reconstruction were carried out.

In November 1970 the Directorate of Transport reviewed the arguments for and against the construction of a trans-island link and decided against it at that time. Their reasons appear to have included doubt that production of farm produce on the western side would be sufficient to justify the relatively high cost; a lingering feeling (based apparently on Public Works inclinations) that improvements to one or other of the west coast surf ports would be adequate for Southwest Bougainville's needs in the immediate future; and a belief that CRA, having already built the Panguna-Jaba River Pump Station road to standards adequate for their own use, would be unwilling either to upgrade that section or to extend their road-building further west. The recommendation of the United Nations Development Programme Transport Survey of 1969 that decision on a trans-island road link should be left until CRA s intentions were clear, and that a trans-island road would be uneconomic in benefit-cost terms, and its apparent favouring of improvements to the Moratona-Buin road and to the port of Kangu also weighed against building the road, as did the Department of Territories decision in September 1967 to restrict road spending on Bougainville to the most important and viable roads, avoiding those based purely on 'political considerations'.

These reasons all undoubtedly appeared valid at the time. In retrospect, however, it can be seen that they underestimated the strength of local feeling in favour of constructing a trans-island road. Events since 1970 may well indicate a widespread recognition by Papuans and New Guineans that in many situations reliance on sea or air transport is now inadequate to meet their needs, and that only surface transport will provide the access they want, despite the risk of interruption by natural causes and the costs of continued maintenance.

In 1970 pressure for a trans-island road from the people of Southwest Bougainville (particularly the northern two-thirds - Nagovisi and Siwai) began to be strongly expressed through Local Government Councils, the elected member for South Bougainville, and the co-operatives. Contributing causes for this included: a recognition that with the approaching end of the construction phase at Panguna BCP might well have equipment available that could be used on road construction projects, and that a request to BCP for such help might be favourably received in view of the Company's often expressed intentions to help the people of Bougainville as much as it reasonably could; the successful construction of the 'Barclays' road' through Siwai, and the goodwill created by the construction company's later help on feeder roads, and its good relations with local people; the increasing volume of cash available in the area from cocoa sales, and the recognition that a trans-island road coupled with the harvesting and sale of maturing cocoa plantings would increase cash incomes even further; and the initiative of the Administration _kiap_ at Boku Patrol Post who recognised that the distance involved was in fact relatively small (about 22 km), and that a low-grade road could be pushed through in part by upgrading already existing tracks, with some mechanical equipment and the help of local people, at a relatively low cost of the order of $50,000.

In 1970 both the Bana and Busiba Co-operatives were advocating the road link. In May 1970 the Bana Producers' Co-operative wrote to the Registrar of Co-operatives, Port Moresby, stressing the need for a transport outlet for the rapidly increasing production of cocoa in Southwest Bougainville in general, and the Banoni-Nagovisi area in particular. The Registrar suggested three solutions, his first preference being for a road link with Kieta via Panguna; the second the hire of a ship to serve south Bougainville, and the third the provision of a permanent ship for the round-Bougainville run; but recognised that the first provided the best solution though the provision of ships would be a temporary palliative while the road was being built.

In June the Busiba Co-operative Association (now in abeyance) wrote to the Co-ordinator of Transport, Port Moresby, arguing for a trans-island road link to join Buin and Kieta. Among other points the letter argued 'the construction of this road would provide access to south Bougainville producers to the Kieta food market which it is hoped they will be entering in the near future. There has been a start made on a beef cattle industry and a road link to Panguna and Kieta would provide possibilities for pigs, poultry, eggs, fresh fruit and vegetables to be supplied to the rapidly expanding population associated with the copper project' (Dept of Transport file 68/73(3) Bougainville District Roads General). The letter also indicated that the amount of cocoa which would be produced in southern Bougainville by 1974-75 would probably represent some 5 per cent of the total Papua New Guinea production, which, if past figures were any indication, would represent approximately 1 per cent of the total Papua New Guinean export earnings.

The DASF agricultural economist working in the field in Southwest Bougainville lent his support to arguments for the road in a report which predicted that cocoa production in Southwest Bougainville could be expected to rise to 2,000 tons by 1972-73. The report emphasised that

> the shipping facilities at Kangu are the major problem faced by the area. There seems little prospect of these facilities being improved, and the societies are faced with heavy stevedoring costs, loss of time by managerial staff, and infrequency of ships. The only satisfactory solution appears to be the construction of a road to link the Kieta-Panguna road to the Nagovisi area to enable south Bougainville's production to be trucked to Kieta for export. (Dept of Transport file 68/73(3).)

This report also pointed out that a trans-island road, which would of necessity pass through inland Nagovisi, would open up one of the most heavily planted cocoa areas of the Boku administrative area not yet served by road.

In September 1970 the road question had apparently been raised again by both BCP and the member for South Bougainville, who were both told by the Administrator that he saw little or no prospect of the early introduction of such an item on the Works Programme (Dept of Transport file 68/73(3)).

By October 1970 the District Commissioner, Bougainville, was also advocating construction of the road: 'Roads exist from Boku to Sovele and from Boku to Moratona...what is in fact required is a road from the pumping station site to Sovele.' He considered that 'putting in a reasonable road from the pumping station site to Sovele would cost between $40,000 and $60,000' and that 'such expenditure was beyond the scope of the Rural Development Fund' (ibid.). In the same letter the District Commissioner reported that 'the matter of a road to link the Buin Subdistrict with Panguna is now without doubt the item being most strongly pressed in this District'. He discussed the difficulties of shipping through Buin, emphasising that they are mainly in the handling of goods from the cargo stores to the vessel. '...the final solution lies in the construction of a road from Panguna to link up with the existing road system in the vicinity of Sovele. I understand that a great deal of local assistance would be forthcoming for this venture' (ibid.).

The strength of local support for the road is shown by a resolution passed in the same month by the Combined District Councils Conference held at Buka. It read:

> That a letter be sent to the District Commissioner, Bougainville, Liaison Officer, Arawa, the Administrator, Prime Minister Gorton, and the Co-ordinator of Transport [asking] that most serious consideration be given to the building of the road between Sovele and the BCP pumping station. The Councils feel that this small section which links up two very important areas of Bougainville is of utmost urgency, and any delay would mean equipment available would be lost. Money for the road is to be repaid to the Australian Government by the Department of Public Works when it is put on the Works Programme (ibid.).

Even this did not produce any immediate result, official thought (especially in the Department of Public Works) still apparently being fixed on the figure of $750,000 for the road. The Patrol Officer at Boku prospected and reported on the possible route in April 1971. His report (Bana L.G.C. Special Report No.1 of 1970-71 (referring to April 1971)) described the first 16 km from Moratona as easy going. The route followed an existing road alignment (which had been purchased by the Administration in 1966) and was already used by tractors in dry weather. The gradient was a gentle 1-2 per cent, and the three main river crossings (the Orei, Nunopa and Pangala) were no worse than those between Buin and Siwai. The route was well-drained, and there were ample quantities of surfacing material conveniently to hand. In the estimation of the Assistant District Commissioner it would be within the capacity of the Bana Local Government Council to upgrade and surface this section of the road, and in fact the Moratona-Sikoreva-Lambalam road was already included in the Bana Council's

1971-72 Rural Development Programme. It was suggested also that other Councils (i.e. Siwai and Buin) and the co-operative societies might be willing to assist the Bana Council in this task on a joint enterprise basis. The remaining 6 km (between Sikoreva and the pumping station) was described as more difficult. The route descended through forest about 500 feet from the top of Sikoreva hill. Road construction would require proper survey and the use of heavy equipment. A bridge would probably be required over the Jaba River near the pumping station, where the river was narrow. It was suggested that BCP could possibly be induced to provide assistance on this section.

At this point individual initiative came into play. Mr Ray Blanchfield drove the Barclays' bulldozer up the existing track from Sovele to Sikoreva and forged a new route down the Sikoreva hill to the pump station. In this enterprise he was supported by the local Nagovisi people, who, for a period of ten days, carried two 44-gallon drums of fuel a day from the Jaba River Pump Station to the bulldozer. It is reported that during this time there were always over 120 men in attendance to carry the fuel (Smith, 1972:16). So officialdom was now presented with something of a _fait accompli_, but much remained to be done before the route could be described as a usable road.

About May 1971 the Bougainville District Co-ordinating Committee called for preliminary examination of the economic potential of the Buin subdistrict, particularly Nagovisi census division, with a view to the conducting of an intensive feasibility study of a road link to connect the Buin-Moratona road with the BCP road terminating at the Jaba River (Dept of Transport file 68/73(3)). The Department of Public Works carried out an engineering survey of the route between the Jaba pump station and Beretemba in 1972 though the route their engineer recommended on Sikoreva hill was not that later used by BCP in the construction of that section. Other engineering studies have also been carried out to find suitable crossing places of the Pangala River. In September 1972 the Department of Public Works made a survey of road needs in the Buin subdistrict (Smith, 1972) which proposed a new set of priorities taking economic changes and the effects of the trans-island road into account.

In June 1971 the Bougainville District Co-ordinating Committee placed the Koro-Sikoreva-Jaba River road first on the Rural Development Programme priority list. In effect this removed the project from a major capital works category with the necessity to find some $750,000 to build a road of moderate standard and converted it to a rural development project requiring funds of the order of $60,000, and resulting in a road of lower standard without major bridges.

It was thought that the lesser amount could be found by pooling the following resources:

(1) Bana Council Rural Development Projects (1971-72):	
(a) Kupon-Taraba Road ($3,700 cash self-help; $3,700 Administration [Rural Development Fund] grant sought	
(b) Taraba-Koro road ($2,000 in kind self-help; $2,000 Administration grant sought	$11,400
(2) Redirection of Rural Development priorities from the Buin and Siwai Councils, probably $4,000, matched by an Administration grant	$8,000
(3) Local donations, probably in excess of $4,000, matched by an Administration grant	$8,000
(4) Other local donations, including co-operative societies, possibly	$4,000
(5) Participation by BCPL, in cash or equipment, possibly	$30,000
	$61,400

Broken down by origin this is:

Indigenous and Council donations	$17,700
Administration RD grant ($5,700 in 1971-72 programme, $8,000 additional)	$13,700
BCP contribution	$30,000
	$61,400

Source: Dept of Transport file 68/73(3).

Unfortunately no conclusive documentary evidence showing how the funds were found is available, or of the precise cost of the road, though it must have been approximately $60,000. Local enthusiasm for the project was running high in 1971, and it received support from both the Administration and BCP. In August 1971 the amount of $10,000 was allocated by the Treasurer from Rural Development funds for the Sikoreva-Jaba road (Dept of Transport file 68/73(3)). BCP also signified their willingness to participate in the road construction to the extent of one-third of the cost. This was to be in the form of making available, at the end of 1971, two bulldozers, one front-end loader and three lorries, and the necessary fuel to run them. The wages and rations of the drivers and of additional labourers that might be required would be provided by the Bana Local

Government Council. The three Local Government Councils concerned (Bana, Siwai and Buin) did contribute; the precise amount of their contribution is not known to me, although it must have been of the order of $10,000. Local Nagovisi people donated both money and labour to at least an equivalent amount. Estimates of the value of the BCP contribution varied according to figures suggested orally to me from $30,000 to $80,000 - depending on whether the source of the estimate was Administration or Company. The Company did donate the used vehicles it had promised, but the Council sold two of them and replaced them with new, smaller units. Difficulties were also experienced with maintenance of the remaining BCP vehicles, and in October 1972 negotiations were continuing to allow the vehicles to be repaired in the BCP workshops at Panguna. The Company constructed a 12.3 m long, 3.75 m wide causeway across the Jaba River near the pump station, and built six km of road over Sikoreva hill, technically the most difficult stretch of the new road. Negotiations were still continuing in October 1972 over the improvement and maintenance of this section, but it appeared likely that in time the standards would be met and maintenance would then become the responsibility of the Department of Public Works. The remaining 16 km of road from Sikoreva to Moratona was built under the auspices of the Bana Council with considerable local help.

By about June 1972 the road was usable although there were immediate problems caused by the severe rains which occurred in the middle of 1972. Most of these had been overcome by October 1972 when the road was easily passable except in times of flood. The earlier floods washed out the section between Kupon and Moratona where the bed of the lower Pangala was being used as a track in October 1972. A new alignment was required there and preparations were underway to construct it. There remains the problem of a permanent crossing over the Pangala River and it seems inevitable that eventually a major bridge will have to be built. The order of cost of this may be about $480,000 at estimated 1975 prices.

The actual construction time of what is at present a low-quality road was not more than one year, and the cost probably approximated $60,000. There will certainly be times when the road is impassable, but it already provides a usable link from the Southwest to Kieta. The amount of use it is already receiving, and the likely increase in this, should ensure that it is gradually improved. In time it will need to be replaced by a better, shorter, trunk road. Naturally the improvement of feeder roads, the possible construction of the Konga-Buin section of the main road through south Bougainville,

Plate 12 Grading the trans-island road near Borioko

Plate 13 Fording the Pangala River on the trans-island road near Borioko, October 1972

and the less immediate but still possible construction of a link to Kieta around the southeast of the island must still be kept under consideration in road planning for the next few years. These matters are further discussed in Chapter 8.

In an attempt to bring more order into future development planning in general and road planning and construction in particular in Southwest Bougainville in the future, a Special Purpose Authority of the three Local Government Councils of the area was being set up in October 1972. Its role would be to establish priorities and initiate and oversee construction of all local road development throughout Southwest Bougainville.

Vehicle ownership

The most immediate effect of the new road will be an increase in vehicle ownership and use in Southwest Bougainville, though this was not yet reflected in official figures by 1972. Table 6.1 and Fig.6.2 show the growth of vehicle registration for Bougainville, by the three registration centres reporting, for the five years 1967-71.

According to the returns published by the Bureau of Statistics, before 1967 there were very few vehicles in the whole island, but with the growth of BCP's operations after that year the numbers show a steep climb to 2,603 in 1971. Most of this took place at the Kieta registry[1] and reflects the arrival of BCP in strength in central Bougainville. The Sohano registry, now serving only northern Bougainville and Buka island, shows a slight increase between 1969 and 1971, while Buin, serving the Southwest, shows only small growth over the same three years.

According to the official source, the total numbers of motor vehicles in Buin subdistrict in 1971 was 128, only 5 per cent of the total for Bougainville. The largest number of these (46) were four-wheel-drive utilities, the type of vehicle best suited to the existing road conditions, particularly the necessity for frequent fording of rivers. There were also 23 tractors and 20 trucks, which again can negotiate rivers and muddy roads. Motor cycles (20) were also fairly popular since they can often negotiate tracks beyond the capabilities of regular vehicles.

---

[1] It is probable that the figures for Kieta do not include additional vehicles used solely on the mine site and registered at Panguna, which does not apparently submit returns to the Bureau of Statistics in Port Moresby.

Table 6.1

Total vehicles registered by types, Bougainville District, 1967-71

Year	Registration centre	Cars	Station wagons	Commercial vehicles: Utilities	Panel vans	Trucks	Other truck types	Omni-buses	Sub-total (c.v.)	Sub-total (c.v. and n.c.v.)	Percentage increase over previous year	Motor cycles	Tractors	Total	Percentage increase over previous year	Trailers or other registrations	Vehicle-equivalent units***	Percentage increase over previous year
1967	Sohano*	8	4	28	-	12	-	-	40	52		20	19	91		18	105.5	
1968	Sohano*	39	13	104	1	51	-	-	156	208	300.0	80	60	348	282.4	53	390	269.7
1969	Buin	6	1	18	-	16	-	-	34	41		14	16	71		6	83	
	Kieta**	60	40	211	-	159	1	6	377	477		62	55	594		35	535.5	
	Sohano	7	3	30	2	20	-	-	52	62		13	21	96		22	121	
Total Bougainville		73	44	259	2	195	1	6	463	580		89	92	761		63	739.5	89.6
1970	Buin	8	-	27	-	16	-	-	43	51	24.4	19	22	92	29.6	11	107	28.9
	Kieta**	164	132	560	7	393	18	21	999	1,295	171.5	97	112	1,504	153.2	35	1,764.5	229.5
	Sohano	12	6	35	1	22	-	1	59	77	24.2	30	23	130	35.4	15	146	20.7
Total Bougainville		184	138	622	8	431	18	22	1,101	1,423		146	157	1,726		61	2,017.5	172.6
1971	Buin	10	8	46	-	20	-	1	67	85	66.7	20	23	128	39.1	11	145	35.5
	Kieta**	370	237	909	7	483	22	38	1,459	2,066	59.5	118	111	2,295	52.6	73	2,629.5	49.0
	Sohano	22	7	46	1	39	-	-	86	115	49.4	22	43	180	38.5	33	226.5	55.1
Total Bougainville		402	252	1,001	8	542	22	39	1,612	2,266		160	177	2,603		117	3,001.0	48.8

* Includes Buin and Kieta.
** It is possible that the figures for Kieta do not include additional vehicles used exclusively on the mine site at Panguna.
*** See Appendix IV.

Source: Bureau of Statistics, Konedobu, Papua New Guinea Statistical Bulletins, Census of Motor Vehicles Registered at 31st December, 1967, 1968, 1969, 1970, 1971.

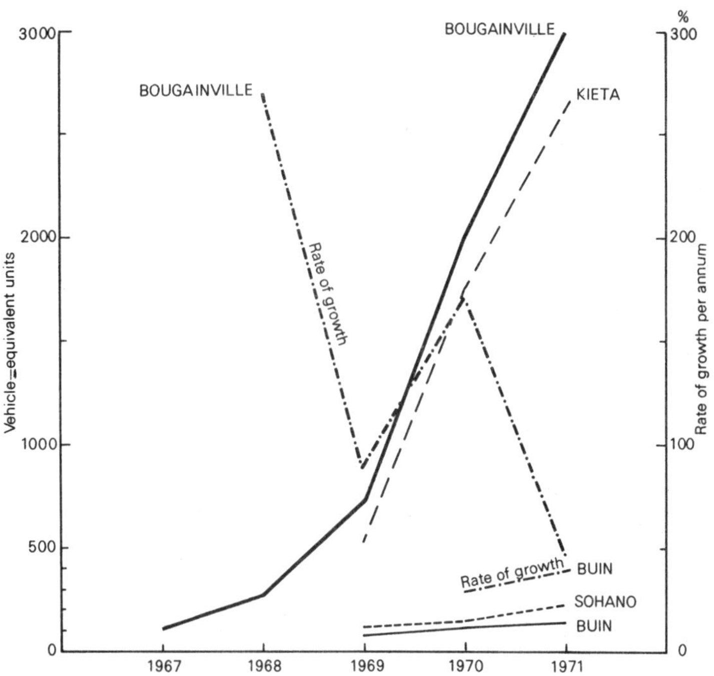

Fig.6.2. Growth of vehicle registration, Bougainville and subdistricts, 1967-71

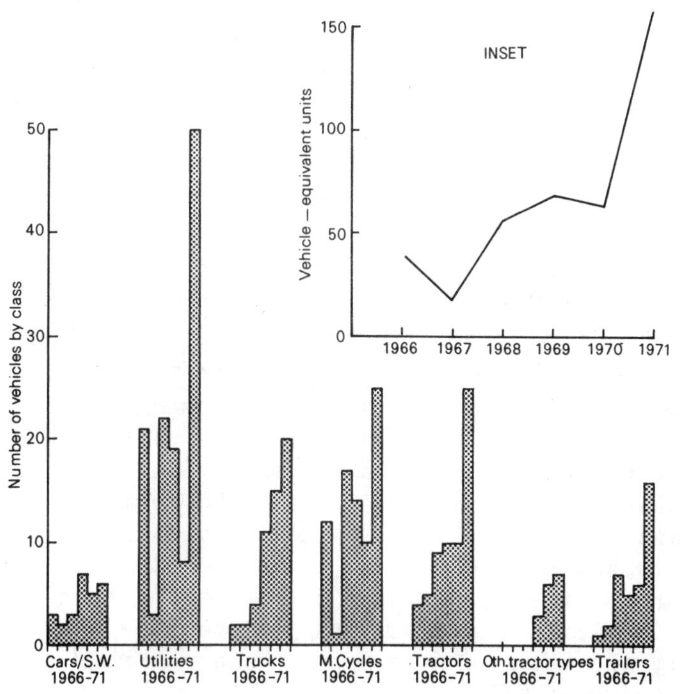

Fig.6.3 Registration of vehicles by types, Southwest Bougainville, 1966-71. Inset: Growth of registrations in vehicle-equivalent units (see Appendix IV)

In an attempt to get fuller data, details of all motor vehicle registrations were collected from the registration duplicates held at the subdistrict office at Buin, and the results are summarised in Table 6.2.

Unfortunately the figures from this source do not entirely agree with the published data, and records for 1967 and 1970 appear to be incomplete, though it may be the case that some 1970 registrations are recorded in the 1971 year. Using this source the total number of vehicles shows a generally steady increase from 1966 to 1971. Figure 6.3 gives a graphical illustration of the importance of various types of vehicles, among which utilities, tractors, trucks and motor cycles are dominant. There was a noticeable increase in registrations of trucks and 'other tractor types' (mainly bulldozers, graders, and front-end loaders) after 1970, due partly to the arrival of heavy road-making equipment for the construction of the Barclays' road in that year.

To facilitate comparisons of total numbers of vehicles an arbitrary equalisation scheme has been set up, based roughly on the weights of vehicles (see Appendix IV). It has been applied to the registrations of vehicles both in the official records (Fig.6.2) and in those obtained from the Buin registry (Fig.6.3 (inset)).

Table 6.3 gives details of ownership of vehicles 1966-71 by Administration, mission and indigenous and non-indigenous owners. In past years the Administration has been the largest vehicle owner in Buin subdistrict, but it has now been largely overtaken by non-indigenous private owners (i.e. European, Chinese and mixed-race businessmen, a plantation owner, and European administrative officers, especially schoolteachers). Mission-owned vehicles have also increased in recent years. Indigenous ownership begins to show a dramatic rise in 1971. It is analysed by vehicle type in Table 6.4.

Some of these vehicles (especially motor cycles) were owned by indigenous Administration officers serving in various departments located in Buin. The tractors and trailers, however, are owned by village people, often local 'big-men', who have sizeable business interests. The rise in registrations of utilities and trucks in 1971 and 1972 represents a response to improved roads on the part of local people who have entered the transport business. It is to be expected that indigenous ownership of utilities and trucks, and perhaps cars or station wagons, will increase as a result of the new road and improved feeder roads.[1]

[1] The patrol officer at Boku predicted that there would be some 300 new vehicles in the Southwest within perhaps a year of the road opening.

Table 6.2

Vehicle registrations, Buin subdistrict, 1966 to September 1972, by vehicle type

	1966	1967*	1968	1969	1970**	1971	1972***
Cars or station wagons	3	2	3	7	5	6	4
Utilities	21	3	22	19	8	50	35
Trucks	2	2	4	11	15	20	11
Motor cycles	12	1	17	14	10	25	6
Tractors	4	5	9	10	10	25	14
Other tractor types	-	-	-	-	3	7	6
Trailers	1	2	7	5	6	16	9
Total	43	15	62	66	57	149	85
Vehicle equivalent units	39.5	17.0	56.5	67.0	64.5	158.0	96.0

* Incomplete. ** Probably incomplete. *** To 30 September.

Source: Vehicle registration duplicates, subdistrict office, Buin.

Table 6.3

Ownership of vehicles, Buin subdistrict, 1966-71

	1966	1967*	1968	1969	1970*	1971	1972**
Administration							
Co-ops and L.G.C.	2	-	6	6	4	8	2
Other Admin.	22	9	12	13	9	41	18
Private							
Mission	3	3	16	6	10	24	9
Non-indigenous	15	3	19	35***	30≠	40	25
Indigenous	1	-	9	6	4	36	31
Total	43	15	62	66	57	149	85

* Incomplete. ** To 30 September.

*** Includes 10 vehicles associated with Barclays' road construction.

≠ Includes 8 vehicles associated with Barclays' road construction.

Source: Vehicle registration duplicates, Buin subdistrict office.

Table 6.4

Indigenous ownership of vehicles, Buin subdistrict, 1966-72, by type of vehicle

	1966	1967*	1968	1969	1970*	1971	1972**
Cars, station wagons	-	-	-	-	1	1	-
Utilities	-	-	3	1	1	18	22
Trucks	-	-	-	-	-	-	2
Motor cycles	1	-	4	3	2	11	1
Tractors	-	-	1	1	-	3	3
Trailers	-	-	1	1	-	3	3
Total	1	-	9	6	4	36	31

* Incomplete ** To 30 September.

Source: Vehicle registration duplicates, Buin subdistrict office.

Figure 6.4 shows the distribution of vehicles owned in Southwest Bougainville in 1971.[1] There is a concentration of Administration and non-indigenously owned vehicles in Buin town, since this is the service centre for the whole area, but there is a sprinkling of indigenously owned vehicles throughout the area served by roads. This pattern can be expected to intensify in the next few years.

Further evidence for increasing participation in transport by local people is found in the licences issued for Passenger Motor Vehicles. Unfortunately data for this are available only from December 1971 to August 1972. During that eight-month period, 28 PMV licences were issued to 27 local people and the Buin Local Government Council to operate PMVs in Southwest Bougainville. Three vehicles were trucks, licensed to carry up to 29, 25 and 32 passengers; the remainder were utilities licensed to carry up to 10 passengers (and a few for 8 or 12). The centres at which these vehicles are based and the passenger capacity in terms of seats available on the routes over which they are licensed to operate are shown in Fig.6.5. The map gives an indication of the areas having the greatest volume of passenger services in 1971-72. With the opening of the new road it is likely that the number of PMV owners will increase and that more will wish to extend their field of operations to include Panguna, Arawa and Kieta.

[1] Buin, Boku Patrol Post and Konga are used as cover addresses by villagers living in nearby census divisions.

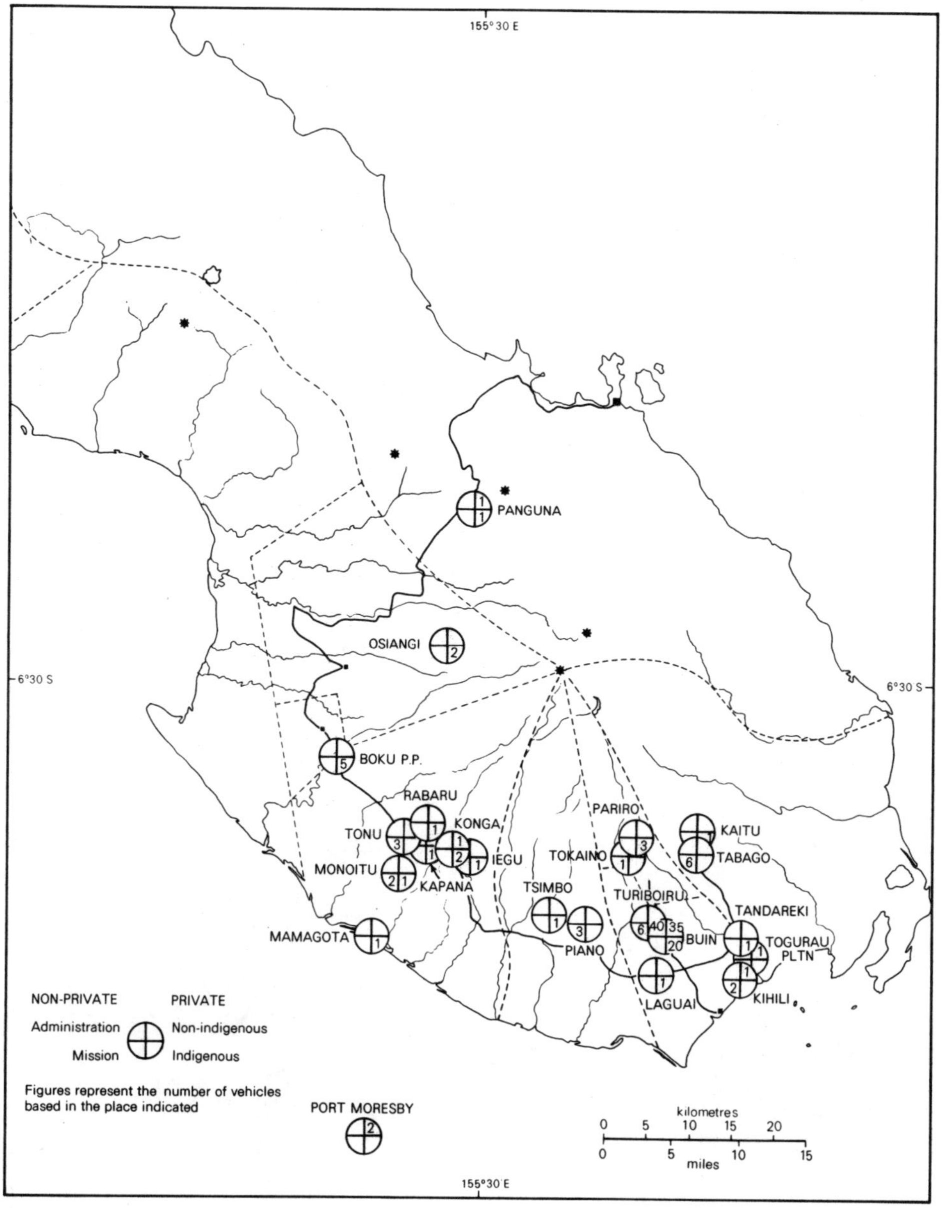

Figure 6.4 Distribution of vehicle registrations, 1971

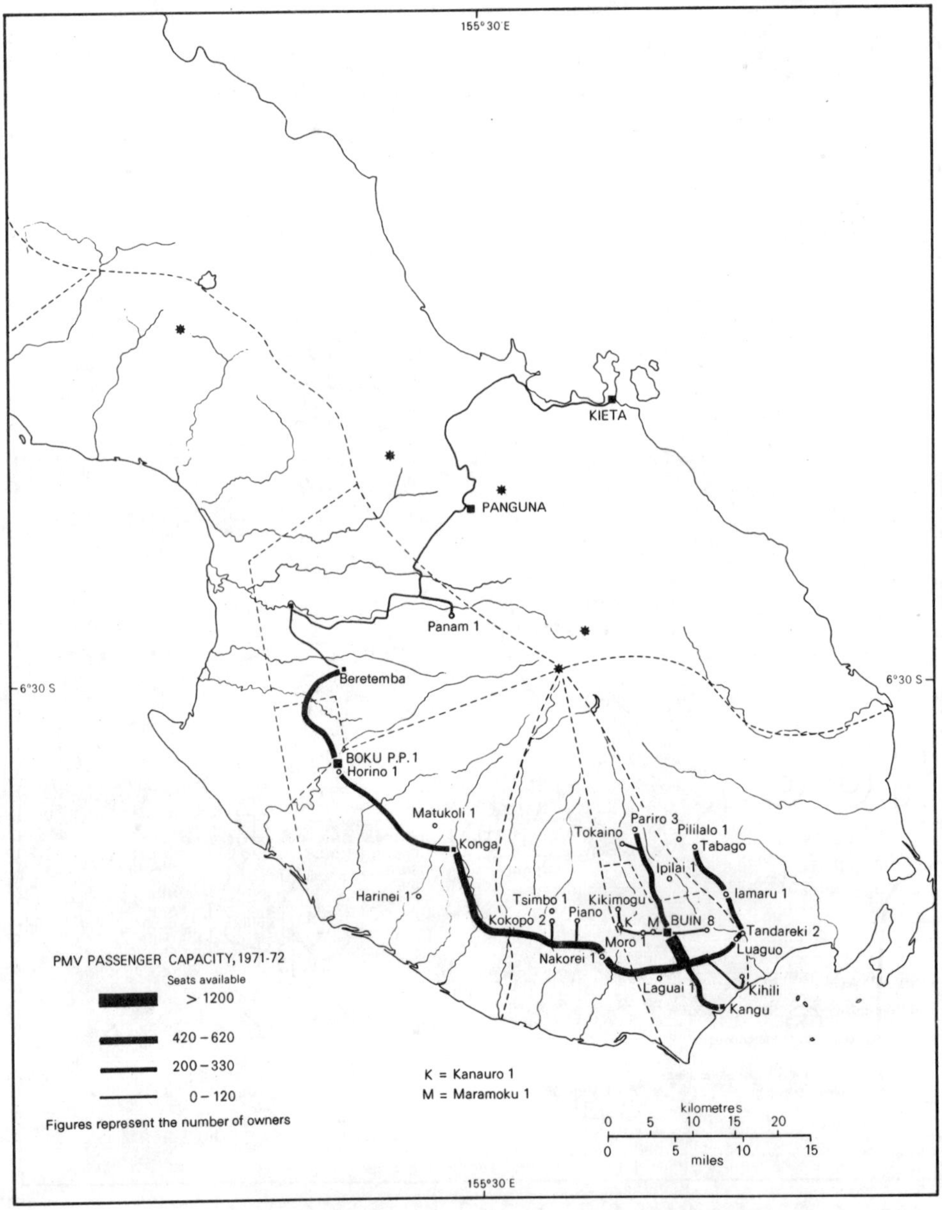

Figure 6.5 Passenger motor vehicle capacity and distribution of PMV ownership, 1971-72

Many more people are of course licensed to drive motor vehicles or ride motor cycles than do in fact own vehicles. Relevant data are given in Tables 6.5, 6.6 and 6.7 and in Figs 6.6 and 6.7. As with numbers of motor vehicles there has been a steep rise in numbers of people holding licences to drive or ride motor vehicles in Bougainville since 1968, and most of this increase is associated with the rise of population at Kieta, Arawa and Panguna. BCP has trained many indigenous drivers and operators of special classes of equipment, especially heavy machines.

Table 6.5

Number of licences to drive a motor vehicle or ride a motor cycle, 1968-71, Bougainville District

Registration centre	1966	1967	1968	1969	1970	1971
	A. Licences to drive					
Buin	n.a.	n.a.	81	121	187	397
Kieta	45	367	506	1,502	3,953	5,930
Sohano	n.a.	n.a.	140	181	420	350
Bougainville	(45)	(367)	727	1,804	4,560	6,677
	B. Licences to ride					
Buin	n.a.	n.a.	29	22	24	15
Kieta	4	82	82	154	225	294
Sohano	n.a.	n.a.	19	26	56	60
Bougainville	(4)	(82)	130	202	305	369

Source: Bureau of Statistics, Statistical Bulletins. Licences issued to drivers and riders of motor vehicles. Years ended 31 December 1967, 1969, 1971.

Licences to drive a motor vehicle issued at Buin begin to show an increase in 1971, and in that year numbered 397, slightly exceeding the figure for Sohano after trailing it in earlier years. Licences issued to ride motor cycles appear to have been declining slightly at Buin; in 1971 there were fewer licences to ride motor vehicles issued than there reputedly were motor cycles in the area. Either people were obtaining their licences elsewhere or had not bothered to renew them.

The race of persons licensed to drive motor vehicles is predominantly non-indigenous at Kieta, somewhat less so at

Table 6.6

Race of persons licensed to drive or ride, 1971

Registration centre	Indigenous No.	Indigenous Per cent	Non-indigenous No.	Non-indigenous Per cent	Total No.	Total Per cent
		A. Licensed to drive				
Buin	165	41.6	232	58.4	397	100.0
Kieta	1,983	33.4	3,947	66.6	5,930	100.0
Sohano	231	66.0	119	34.0	350	100.0
Bougainville	2,379	35.6	4,298	64.4	6,677	100.0
		B. Licensed to ride				
Buin	9	60.0	6	40.0	15	100.0
Kieta	62	21.1	232	78.9	294	100.0
Sohano	29	48.3	31	51.7	60	100.0
Bougainville	100	27.1	269	72.9	369	100.0

Source: as for Table 6.5.

Table 6.7

Sex of persons licensed to drive or ride, 1971

Registration centre	Males No.	Males Per cent	Females No.	Females Per cent	Persons No.	Persons Per cent
		A. Licensed to drive				
Buin	374	94.2	23	5.8	397	100.0
Kieta	5,518	93.1	412	6.9	5,930	100.0
Sohano	312	89.1	38	10.9	350	100.0
Bougainville	6,204	92.9	473	7.1	6,677	100.0
		B. Licensed to ride				
Buin	14	93.3	1	6.7	15	100.0
Kieta	286	97.3	8	2.7	294	100.0
Sohano	58	96.7	2	3.3	60	100.0
Bougainville	358	97.0	11	3.0	369	100.0

Source: as for Table 6.5.

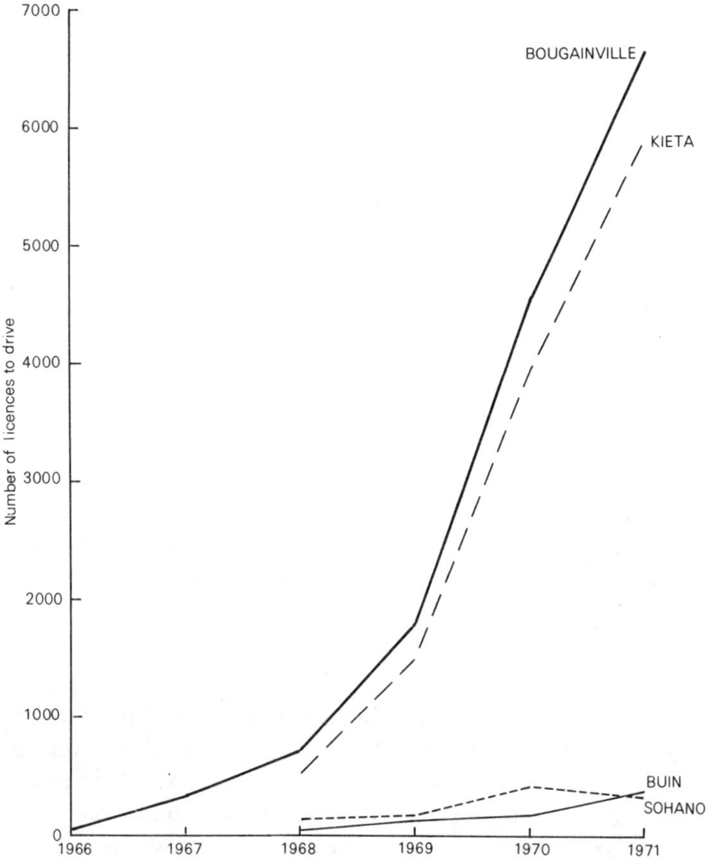

Figure 6.6 Numbers of licences to drive, Bougainville and sub-districts, 1966-71

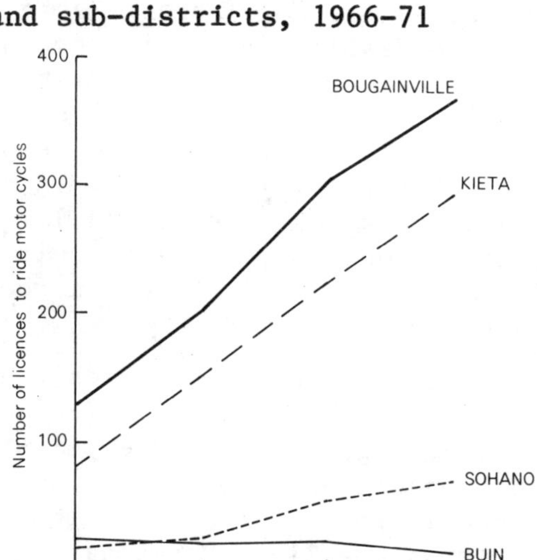

Figure 6.7 Numbers of licences to ride motor cycles, Bougainville and sub-districts, 1968-71

Buin, and predominantly indigenous at Sohano. It is to be expected that the numbers of indigenous people holding drivers' licences issued at Buin will increase more rapidly in the next few years than the numbers of non-indigenous licence holders. In the case of licences to ride motor cycles non-indigenous holders predominate at Kieta and Sohano, while indigenous holders predominate in the small numbers issued at Buin. Males predominate strongly over females among holders of drivers' or riders' licences at all registration centres, the ratio being 9:1 in all cases.

Future vehicle numbers

With the opening of the new road and improvements in feeder roads, vehicle ownership, especially by indigenous people, will increase in the next few years. This is likely to occur first in utilities, tractors and trucks. The rise in numbers of utilities registered in Buin in 1971, even if some are attributed to 1970, is probably the beginning of this phenomenon.

Predictions about future numbers of vehicles in the area are likely to be inaccurate. The number of Europeans or other non-indigenous people living permanently in Southwest Bougainville is not likely to increase greatly in the future, unless major construction works occur again, as would temporarily be the case if the Konga-Buin or Buin-Kieta roads were to be built, or if the timber scheme at Tonolei Harbour were to be reactivated. The present work needs of the Administration are largely met by its present number of vehicles. There is some prospect that private non-indigenous vehicle ownership may increase slightly if residents feel that the new access to Kieta warrants private ownership. Non-indigenous people in commerce may wish to increase the number of vehicles they own, to keep pace with increasing commercial activity generally. In neither case do there appear to be grounds for expecting a large increase in non-indigenous vehicle ownership. Nor does it appear likely that there will be a very large increase in the number of vehicles owned by the missions.

Indigenous ownership and use of vehicles can, however, be expected to increase rapidly in the next few years, probably levelling out after perhaps two or three years, though this will depend as much on the price of cocoa as on other factors such as improved feeder roads. When the Rigo road (which is locationally similar to the Kieta-Moratona link) was opened in 1965 there was an increase in all motor vehicles of 104 per cent in the first year and 23 per cent in the second (Ward, 1970:28). It is impossible to predict whether similar rates will apply in Southwest Bougainville, but the general sense of economic optimism prevailing in the area in October 1972 would seem to indicate a high rather than low response to the new opportunities offered by the road. Table 6.8 gives high and low projections for total numbers of motor vehicles in Southwest Bougainville for each year from 1971 to 1982. For the high projection rates of 100 per cent for the first year and 23 per cent for the second year, followed by growth at the national average rate of 14.8 per cent for successive years were used. For the low projection more conservative rates of 50 per cent and 20 per cent for the first two years respectively were used, followed by 14.8 per cent for the remainder of the period.

Table 6.8

High and low projections of total motor vehicle numbers in Southwest Bougainville, 1971-82

Year	High projection			Low projection		
	Total numbers	Rate % p.a.	Increase per year	Total numbers	Rate % p.a.	Increase per year
1971	128	100	128	128	50	64
1972	256	23	59	192	20	38
1973	315	14.8	47	230	14.8	34
1974	362	14.8	54	264	14.8	39
1975	416	14.8	62	303	14.8	45
1976	478	14.8	71	348	14.8	52
1977	549	14.8	81	400	14.8	59
1978	630	14.8	93	459	14.8	68
1979	723	14.8	107	527	14.8	78
1980	830	14.8	123	605	14.8	90
1981	953	14.8	141	695	14.8	103
1982	1,094			798		

Assuming no major external or internal economic changes, it is possible therefore that by 1982 there will be between 800 and 1,100 vehicles owned in Southwest Bougainville. Using the projected populations for 1982 (Table 2.7) this would imply one vehicle for every 40-60 persons, which is by no means saturation point. In 1971-72 there was one vehicle for every 243 persons.

Traffic

The numbers of vehicles travelling on the roads of Southwest Bougainville will in future be greater than the number actually owned there, since vehicles from Kieta, Panguna and Arawa will use the new trans-island road to come into the area.

So far as I am aware there were no official traffic counts on Southwest Bougainville roads before 1972. Estimates of 1972 traffic and predictions for 1982 traffic are, however, given by Smith (1972:18a). Since the information given in his table is relevant to the present study it is reproduced in full here (Table 6.9).

My observations of traffic on various sections of the road in dry weather in October 1972 tend to confirm his estimates, though mine were incomplete. A further partial estimate of traffic using the main roads of Southwest Bougainville is made by Couper and Cotter who predict that by 1980 to meet the cargo carrying needs of the area 8 trucks per day will be required between Buin and Konga, 11 trucks per day between Konga and Beretemba, and 14 trucks per day between Beretemba and Kieta (Couper and Cotter, 1972:104). These estimates are consistent with Smith's predictions. Traffic volumes on the major feeder roads making up the south Bougainville network are likely to be less than on the main road, and to decrease the greater the distance from the main road. Smith's 1972 and 1982 predictions have been mapped in Figs 6.8 and 6.9.

The only available data for traffic flows derive from approximate counts made under the direction of the Patrol Officer, Boku, near Koro village on the Mokokolio-Jaba River section of the new road. They are: for July-August 1972 about 20 vehicles daily average, and for September-October 1972 about 40 vehicles weekday average and up to 80 vehicles per day at weekends (pers. comm., P.O. Boku, October 1972).

A large rise in traffic at weekends was also found on the Rigo road (Ward, 1970:36), and on the Hiritano Highway where it is almost certainly a consequence of Port Moresby residents using the road for recreational purposes on Saturdays and Sundays. Bouchard (1972:97-102) did not find a similar pattern on the Okapa road, although his figures are for an incomplete week in one case, and are obscured by an atypical event during the second week. This road, however, does not offer the same attractions for recreational use by outsiders to the area at weekends as do the Hiritano Highway, the Rigo road and the Bougainville trans-island road. It is to be expected that a weekend rise in traffic will continue to be experienced on the Bougainville trans-island road, as residents of Panguna, Arawa and Kieta take the opportunity to visit Southwest Bougainville. Buin market will presumably continue to be held on Saturdays, thereby generating traffic in the southern part of the area.

Table 6.9

Standards, estimated traffic 1972 and 1982, life and maintenance costs of Bougainville trans-island road, prepared by Smith (1972)

Bougainville trans-island road	Km	Minutes	Average engineering standard	Average formation width m	Average pavement width m	Average pavement thickness	1972 average (dry weather) vehicles/day (estimate)	1982 average (dry weather) vehicles/day	Estimated life of road with existing maintenance and expected traffic	Existing annual maintenance expenditure/mile	Total
Buin-Konga	39	65	Rural III Flat	4.8	2.4	0.25m sand	30-40	75-100	10 years	1,000	25,000
Konga-Boku	32	35	Rural II Flat/Undul.	7.2	3.6	0.38m	30-40	75-100	15 years	1,000	23,000
Boku-Beretemba	18	25	Feeder Undul./Hilly	4.8	3.0	0.10m sand	30-40	75-100	2-3 years	(2,000)	(22,000)
Beretemba-Moratona	11	20	Feeder Undul./Hilly	4.8	2.4	(currently being reconstructed)	30-40	100-150 (if main road)	2-3 years	Current reconstruction (2,000)	(14,000)
Moratona-Mokokolio	16	50	Feeder Undulating	3.6	2.4	0.08m sand (currently being reconstructed)	20-30	100-150 (if main road)	6 months	(2,000) (current reconstruction)	(20,000)
Mokokolio-Jaba River	10	30	Feeder Mountainous	7.2	3.6	0.15m river boulders on 3.4 km Nil on 6.2 km	10-20	125-175	6 months	Nil	Nil
Jaba River-Panguna	10	20	Rural II Mountainous	7.2	5.4	0.46m river gravel	50-100	150-200	15 years	n.a.	n.a.
Panguna-Kieta	37	65	Rural I and II Mountainous/ Flat	11.4	9.0	0.61m crushed gravel and rock	2,000-3,000	3,000-5,000	20 years	n.a.	n.a.
	173	310									

Source: Smith (1972:18a). (Original in miles and feet.)

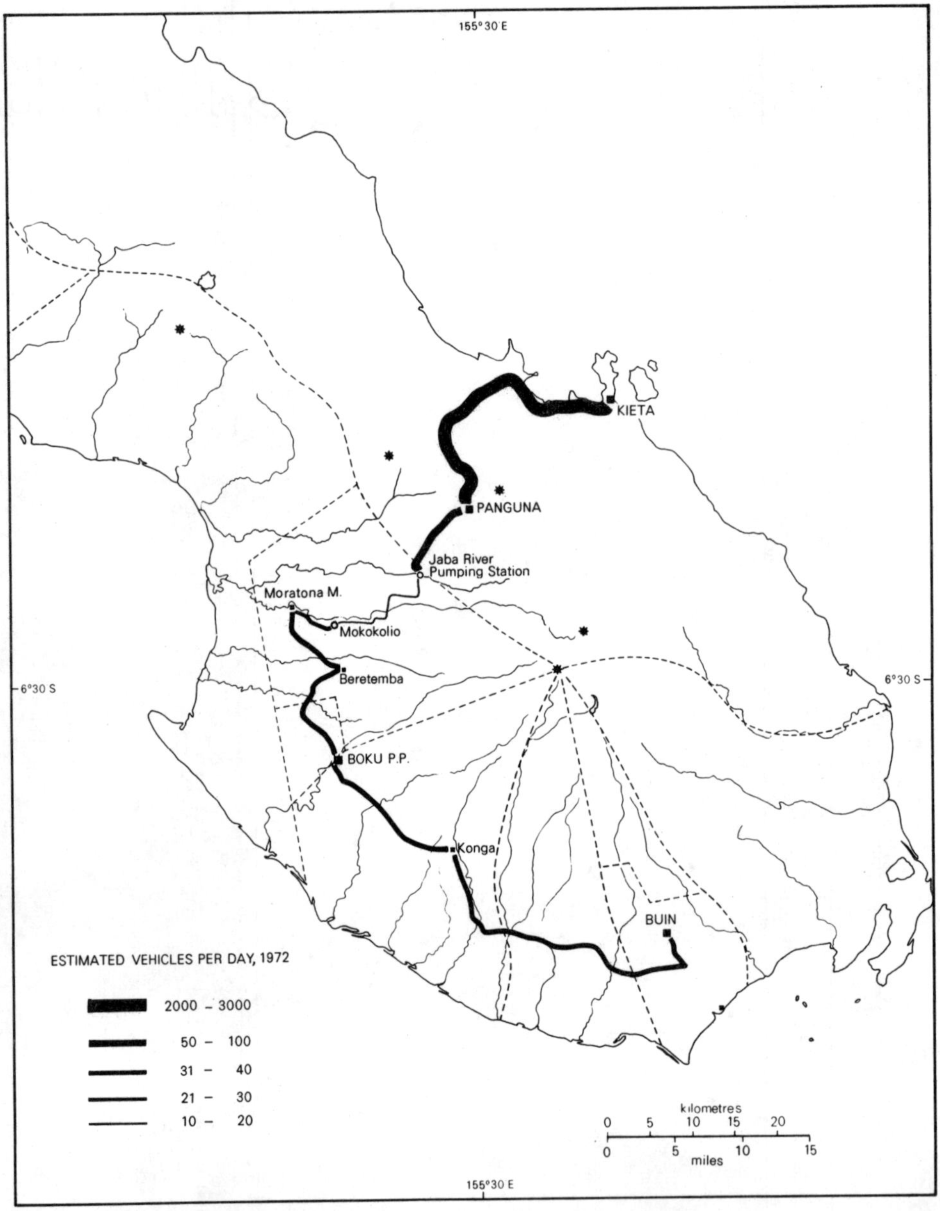

Figure 6.8 Estimated vehicles per day, 1972

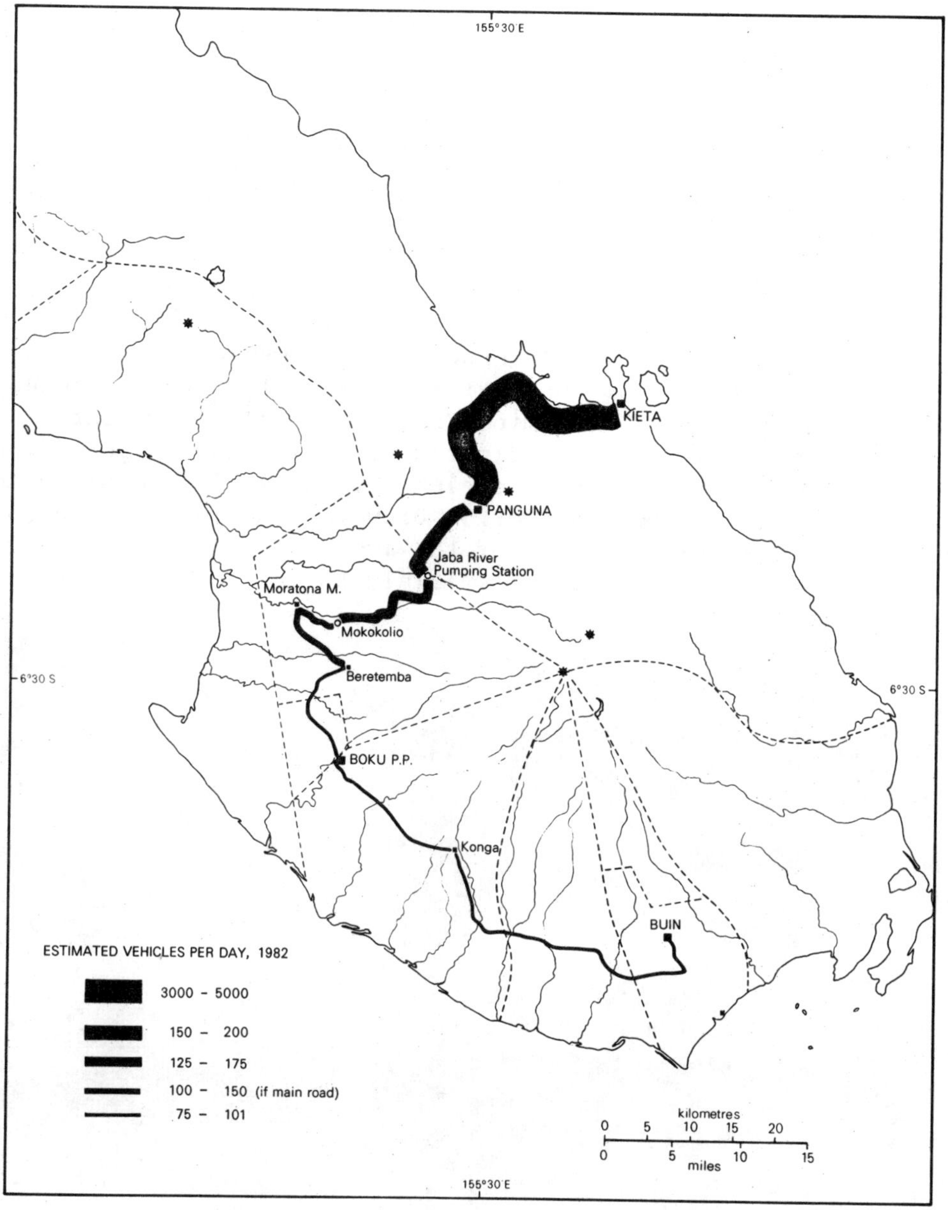

Figure 6.9 Estimated vehicles per day, 1982

There is a small market at Konga on Saturdays which is likely to have a similar though smaller effect in the central part of Southwest Bougainville. BCP's vegetable buying market at Panguna on Fridays is likely to generate traffic in northern and eastern Nagovisi on that day.

The composition of weekday traffic in Southwest Bougainville is likely to remain heavily biased towards utilities, trucks, tractors and trailers for some years, at least while the necessity to ford rivers and negotiate poor quality feeder roads remains. These are the working vehicles which carry wet cocoa beans to the fermentaries. Larger trucks will be used increasingly for long-distance movements carrying dry cocoa to Kieta and Kangu for export, and consumer goods into Southwest Bougainville. Utilities are also favoured vehicles for passenger carriage. Administration, Local Government Council and Co-operatives business travel is also largely done in utilities, although motor cycles have a use in reaching the extremities of the road network. Bicycles should continue to be used in the flatter parts of the area. At weekends the private car or station wagon is likely to make up a higher proportion of traffic.

Chapter 7

Existing transport systems: air transport, coastal shipping, freight charges and transport costs

Air transport

Bougainville District has not experienced the strong dependence on small airstrips served by light aircraft which has characterised the development of many areas in mainland Papua New Guinea. Its island nature and the relatively early date of its initial commercial development meant that its transport services were oriented towards coastal shipping, which could readily provide access to coastal plantations and offshore islands. There are only nineteen airstrips in the District, sixteen on Bougainville Island, with two others (including one at Moratona in the Southwest) having been recently closed, and one other due to open shortly.

In the Southwest in 1972 there were seven airfields (shown in Fig.7.1 and described in Table 7.1). Buin airstrip was used by Fokker Friendships until early 1972, with four scheduled services from Kieta to Buin per week by TAA and Ansett. Poor drainage caused considerable quantities of water to lie on the surface of the airfield during and after rain, making its use hazardous. It is probably for this reason that it has now been reduced to DC3 level.

Smaller airfields capable of handling light aircraft only were located at Boku, Tonu, Oria, Lehu, Torokina, and Siangaru. Torokina was rehabilitated from the main wartime airstrip there and reopened in 1972 to serve the limited transport needs of its area. An airfield was completed by villagers at Oria in 1971, to provide a means of exporting the cocoa and vegetables produced in this otherwise inaccessible upland area.

Details of aircraft movements, passengers, freight and mail carried at the airfields of Southwest Bougainville are given in Table 7.2. The table shows that Buin handled a significant amount of passengers and freight (although the amount of freight is insignificant when compared to the amount moved by

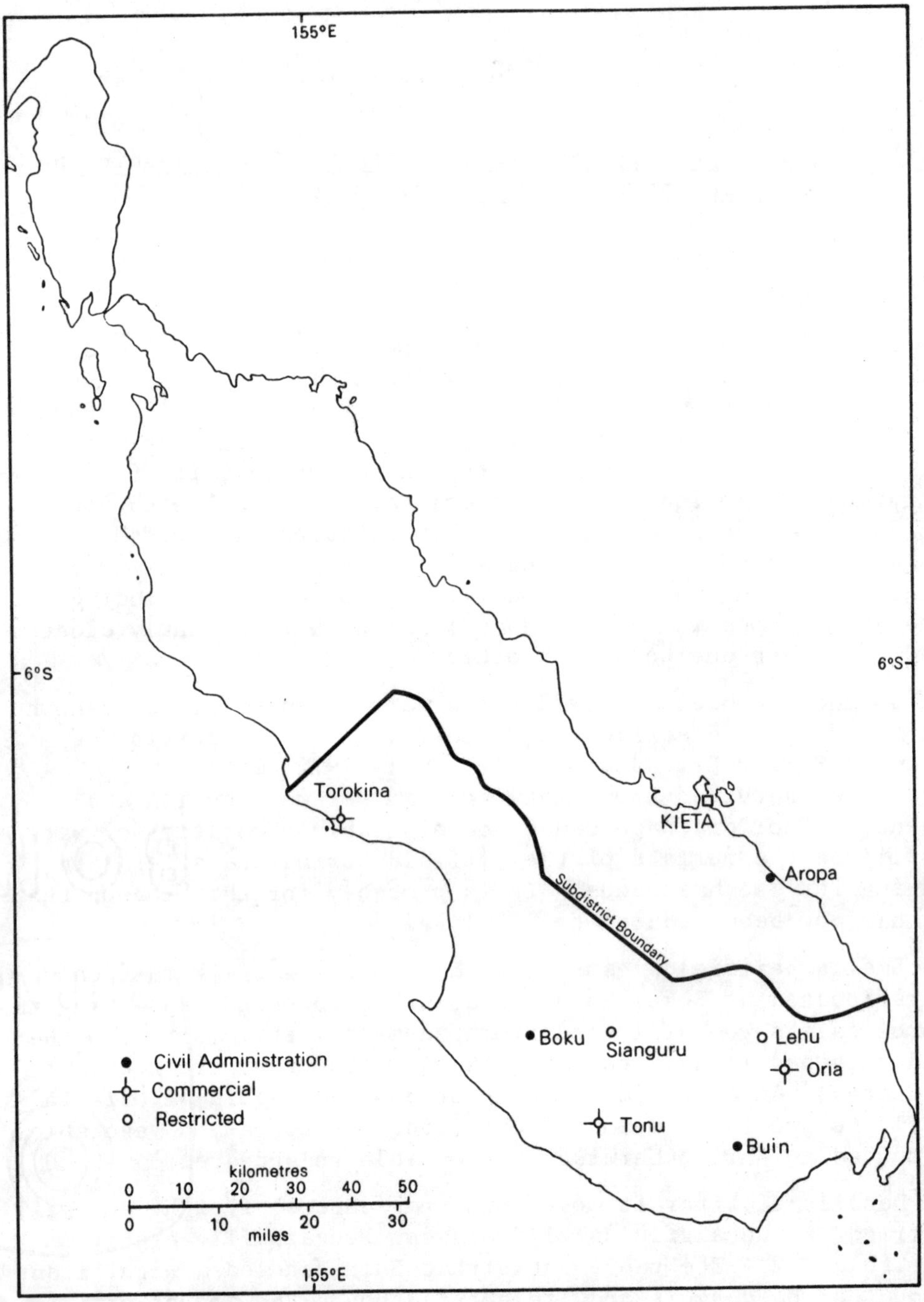

Figure 7.1 Airfields

Table 7.1

Airfields in Southwest Bougainville, 1972

Place	Date opened	Latitude/Longitude		Owner	Class*	Dimensions (m)	Aircraft type
Buin	Before 1965	06°43'	155°42'	Civil Administration	X	1,650x60	DC3
Torokina	21.7.72	06°13'	155°03'	Bana L.G.C.	ALA	840x45 supp. (45)	Baron
Boku	26.9.68	06°32'	155°21'	Civil Administration	Y	750x60	C402
Tonu	Restricted 27.6.66 Commercial 13.5.71	06°41'	155°27'	United Church	ALA	624x30 supp. (45)	BN2A
Oria	24.6.71	06°36'	155°46'	People of Oria village	ALA	600x30 supp. (45)	BN2A
Siangaru	Commercial 29.9.67 Restricted 17.4.72	06°32'	155°29'	United Church	ALA	450x30 supp. (45)	C185
Lehu	5.9.69	06°33'	155°44'	M.A.F.	ALA	450x30 supp. (45)	C185
Moratona	1.8.66, closed 20.10.72	06°27'	155°51'	Catholic Mission	ALA	450x30 supp. (45)	-

* ALA (Authorised Landing area): an aerodrome suitable for operations by aircraft having maximum take-off weights not exceeding 5.7 tonnes engaged in charter, aerial work and private operations, but excluding Regular Public Transport operations.
Class X: an aerodrome suitable for operations by aircraft having maximum take-off weights not exceeding 22.7 tonnes but excluding the Fokker Friendship.
Class Y: an aerodrome suitable for operations by aircraft having maximum take-off weights not exceeding 5.7 tonnes engaged in regular public transport, charter, aerial work and private operations.

Source: DCA, Port Moresby, May 1973. (Original in pounds and feet.)

Table 7.2

Air movements, passengers, freight and mail carried at airfields of Southwest Bougainville, 1972

	Buin*	Boku**	Tonu**	Oria**	Lehu**	Siangaru**
1971						
Aircraft movements***	860	261	232	-	2	-
Passengers disembarked	7,952	207	277	-	-	-
Passengers embarked		290	308	-	-	-
Freight disembarked (kg)	31,518 (=32 tonnes)	320 (=0.3 tonnes)	264 (=0.3 tonnes)	-	-	-
Freight embarked (kg)		3	0.5	-	272	-
Mail disembarked (kg)	-	-	-	-	-	-
Mail embarked (kg)	9	-	-	-	-	-
1972						
Aircraft movements***	2,286	433	1,209	755	99	54
Passengers disembarked	10,321	455	1,254	777	118	60
Passengers embarked		356	1,332	689	108	58
Freight disembarked (kg)	65,136 (=65 tonnes)	8,921 (=9 tonnes)	4,852 (=5 tonnes)	17,566 (18 tonnes)	2,734 (=3 tonnes)	-
Freight embarked (kg)		3,305 (=3 tonnes)	530 (=0.5 tonnes)	45,211 (=45 tonnes)	354 (=0.4 tonnes)	-
Mail disembarked (kg)	-	-	57	-	82	-
Mail embarked (kg)	-	-	-	-	-	-

* Includes regular public transport services, airline charter, and charter by all small operators.
** Refers to charter by all small operators and possibly some airline charter; there were no regular public transport services to these airfields.
*** Each take-off or landing counts as one movement.

Source: DCA, Port Moresby, May 1973. (Original in pounds.)

sea through Kangu). Those amounts have been increasing in recent years as comparison with Couper and Cotter's data shows.

Table 7.3

Passengers and freight (in and out) through Buin airport, 1969-72

Year	Passengers		Freight (tonnes)	
	Couper and Cotter	DCA	Couper and Cotter	DCA
1969	3,971	n.a.	35	n.a.
1970	5,426	n.a.	39	n.a.
1971	8,702	7,952	55	32
1972	-	10,321	-	65

Source: Couper and Cotter, 1972:72 and DCA, Port Moresby, May 1973. (Originals in tons.)

Apart from Buin only Oria airfield handled sizeable quantities of freight in 1972. The outwards freight consisted of cocoa and fresh fruit and vegetables, while inward freight was presumably store goods. Table 7.4 gives anticipated cocoa production for the Oria area 1971-75. The air freight on cocoa from Oria to Kieta is 2 cents/lb (equivalent to 0.91 cents/kg or $45.72 per tonne) without backload, but this is reduced by backloading which occurs most of the time (pers. comm., Manager, Bougainville Air Services, Kieta, October 1972).

Table 7.4

Projected Oria area cocoa production, 1971-75

Year	Anticipated* production		Revenue**	Air*** freight	Return to grower
	bags/wk	tonnes/yr	$	$	$
1971	11	37.4	18,400	1,656	16,744
1972	13	44.2	21,750	1,958	19,792
1973	17	59.1	29,100	2,619	26,481
1974	20	67.1	33,000	2,970	30,030
1975	25	86.2	42,400	3,816	38,584

* Based on 40% recovery rate from wet bean and 2.7 kg tree yield.

** Based on standard price of $500/ton (i.e. $508/tonne).

*** At $45/ton (i.e. $45.72/tonne).

Source: Production figures supplied by DASF and Department of Trade and Industry.

Plate 14. Oria airstrip, Kono census division

In addition to cocoa the Oria area produces vegetables and fruit, which in 1972 were being flown out at an average rate of about 272 kg per week (Couper and Cotter, 1972:73). The freight is again 2 cents/lb (0.91 cents/kg). Some of this produce is sold to the hotels and hospital in Kieta and Arawa, and the rest used privately by Bougainville Air Services personnel.

There is another small airstrip at Lehu about four miles northwest of Oria, from which vegetables are also flown out. In 1972 about 182 kg per week were being uplifted. Bougainville Air Services provided the transport free but paid villagers $10 for an average load, i.e. 2.5 cents/lb (i.e. 1.14 cents/kg) (pers. comm., Manager, Bougainville Air Services, October 1972). Although BCP has shown interest in the fruit and vegetables produced in the Oria area it was not buying from that area in 1972. There is a large still unsatisfied market for fresh fruit and vegetables (particularly for quality high altitude vegetables) in the urban complex of Kieta-Arawa-Panguna, and the prospects for extending this economic activity are good if the villagers want to undertake it.

Air transport in Soughwest Bougainville now has two aspects of importance: the transport of passengers in and out of Buin, and of cocoa and vegetables from Oria. The opening of the trans-island road may well reduce the numbers of passengers carried to and from Buin since passenger fares by road will, when established, undoubtedly be cheaper than air fares. In addition, passengers will be able to travel by road much nearer to their destination if this is not Buin. Air services will, however, almost certainly need to be maintained between Kieta and Buin to provide rapid communications and possibly as a tourist service in future.

The trans-island road will have no effect on the Oria area, which has little immediate prospect of a road connection to Buin or Kieta.

Coastal shipping

Before the trans-island road was opened Southwest Bougainville depended almost entirely on the port of Kangu for its exports of cocoa and copra and its imports.

Kangu is on the south coast some seven miles by good road from Buin town. The coast is exposed to the southeast winds of May to November, when heavy surf may make it impossible to transfer cargo between ship and shore. There is a rudimentary breakwater and jetty built in 1958-59 and extended seawards in 1966 and 1969. Although this structure may have formerly afforded some shelter to small vessels, its construction has resulted in the deposition of sand by very active longshore drift on both sides of the jetty, making it necessary to go 75 m from the end of the jetty to reach 3.9 m of water (Couper and Cotter, 1972:21). Small coastal vessels must anchor 45-150 m off and use surf boats to convey cargo to and from the shore. The Shell tanker *Sigiwai* while at anchor uses 120 m of hose to discharge (loc. cit.). The risk of delay and damage to cargoes is consequently high. Shore facilities consist of three sound storage sheds 18 m by 6 m. There is a light railway track along the jetty but it is not apparently used, and all cargo is transferred between beach and sheds by hand labour (Couper and Cotter, 1972:43).

During 1971 Kangu received about ten calls per month. The M.V. *Nukumanu* of 264 NRT operated by Steamships Trading Co. called every 3 weeks; the M.V. *Waiotahi* of 103 NRT (owned by Tapaka Co., agents Bismark Shipping Co.) called every 10 days; and the *Ninsa II* of 98 NRT (operated by A. and E. Tong) called every 8 days. In addition a work boat from Kieta called every 3-4 days, and mission craft called occasionally (Couper and Cotter, 1972:24-5, 28).

Table 7.5

Imports and exports through Kangu, August 1971-September 1972
(tons)

Period	General cargo by weight	General cargo by volume*	Timber**	Rice***	Fuel≠	Vehicles≠≠	Total
			Imports				
August-December 1971	376	1,245	98	62	355	148	2,284
Extrapolated to full year							4,482
January-September 1972	307	1,384	50	125	325	472	2,663
Extrapolated to full year							3,551
October 1971-September 1972	392	1,738	87	416	432	508	3,573

Period	Cocoa≠≠≠	Copra∅	Coffee∅∅	Empty drums≠	Vehicles≠≠	Total
			Exports			
August-December 1971	978	85	2	354	38	1,457
Extrapolated to full year						3,497
January-September 1972	842	64	-	302	11	1,219
Extrapolated to full year						1,625

* Converted at 40 cu.ft = 1 ton
** Converted at 480 super ft = 1 ton
*** Converted at 40 bags = 1 ton
∅ Converted at 14.5 bags = 1 ton
≠ Converted at 3x44 gal. drums = 1 ton
≠≠ Converted at 40 cu.ft = 1 ton
≠≠≠ Converted at 16 bags = 1 ton
∅∅ Converted at 17 bags = 1 ton

Source: Buin Producers' Co-operative Ltd Stevedoring Agents, Buin.

The amounts of cargo handled at Kangu during the period August 1971-September 1972 are given in Table 7.5. These figures are comparable with those found by Couper and Cotter (1972:48) which are given in Table 7.6. Couper and Cotter also give additional data for exports of cocoa and copra through Kangu by destination. These are given in Table 7.7.

Table 7.6

Cargo (in revenue tons) unloaded (A) and loaded (B) at Kangu

Category	1968-69	1969-70	1970-71	1971-72 (8 mths)	1971-72 (extrapolated to full year)
(A) Inward cargo					
General	-	-	-	2,256	3,384
Drums of fuel	-	-	-	297	446
Rice	-	-	-	206	309
Timber	-	-	-	120	180
Vehicles	-	-	-	23	35
Total p.a.	1,250	3,300	3,910	2,902	4,353

	1968-69	1970-71 (estimated)	1971-72 (extrapolated)
(B) Outward cargo			
Cocoa	260	936	1,800
Copra	105	117	140
Other	150	150	140
Total p.a.	515	1,203	2,080

Source: Couper and Cotter (1972:48) citing UNDP E6.2; Capt. W. Gibson's Report, Ref. D.T. 68/373, 1971-72; Co-op. Stevedoring Records, Buin.

These data produce a total of 1,358 tons for 9 months which can be extrapolated to 1,811 tons for a full year. In addition production of copra in Buin subdistrict is given as 150 tons, all of which would have been exported through Kangu. It appears likely, therefore, that around 1971-72 Kangu was importing about 4,000 tons of cargo and exporting about 2,000 tons. Until around 1970-71 most of Kangu's trade was with Rabaul, but with improved shipping facilities at Kieta, that port has taken over Rabaul's position as the superior port for Kangu.

Table 7.7

Cocoa exported through Kangu, 1 June 1971-29 February 1972, i.e. 9 months

	Kieta			Rabaul	
	Bags	Tons	%	Bags	Tons
Bana Society	4,050	253	22	-	-
Siwai Society	3,579	224	19	-	-
Buin Society	7,626	477	41	-	-
Siwai/Buin non-co-op.	3,456	216	18	-	-
Kangu	18,711	1,170	100	c.3,000	188
Total: 1,358 tons					

Source: Couper and Cotter (1972): Figure 6(b), citing Produce Inspector, Kieta.

Table 7.8

Stevedoring charges at Kangu, 1972

Class	Charge	Equivalent charge per ton weight
		$
1. By weight	$2.40	2.40
2. By volume	5 cents per cubic foot	2.00
3. Timber	4 cents per super foot	1.20
4. Bags of produce		
Cocoa	10 cents per bag	1.60
Copra	10 cents per bag	1.45
Coffee	10 cents per bag	1.70
Rice	6 cents per bag	2.40
5. 44 gallon drums		
Full	40 cents per drum	1.20
Empty	10 cents per drum	0.30
6. Vehicles	2 cents per cubic foot	0.80

Source: Buin Producers' Co-operative Ltd.

All cargo handling at Kangu port is done by the Buin Producers' Co-operative who are the stevedoring agents for the port. Workers are not trained for the job, being simply called from their villages when required. Consequently, as Couper and Cotter point out (1972:81) their unfamiliarity with the proper methods of carrying out the work, together with the difficult conditions, make for a relatively large amount of cargo damage at Kangu. This is likely to contribute to a change to road transport by shippers and consignees in Southwest Bougainville. Stevedoring charges at Kangu in 1972 are given in Table 7.8.

It should be noted that most general cargo passing through Kangu is charged on a measurement basis, but cocoa and copra, despite being measurement cargoes are charged by weight. Hence cocoa shippers in Southwest Bougainville gain a saving of some 80 cents per ton on steverdoring costs by being charged by weight rather than measurement. A similar saving accrues to shippers of cocoa and copra on sea freight rates (Couper and Cotter, 1972:52).

Freight charges and transport costs

Freight charges by sea from Kangu to Kieta were variously quoted as \$12.50, \$11.50, or \$11.00 per ton. From Kangu to Rabaul they were \$18.50, \$17.50, or \$15.50 a ton (Couper and Cotter, 1972:53). In addition a shipper must meet charges of \$1.80 per ton for wharfage and handling in Kieta; 20c per ton cartage ship to shore; and \$3.00 per ton insurance (B.F. Darcey, pers. comm.). Thus transport costs from Kangu to instore Kieta total \$16.50 per ton (using \$11.50 as the sea freight charge). In addition there are road freight charges to Kangu from the inland origin. These are \$30/load (i.e. \$6 per ton) from Konga to Buin (pers. comm., Secretary, Motuna Co-op. Wholesale Society, Kangu) and \$70 per load (\$14/ton) from Beretemba to Buin. On a per ton (one way) basis Couper and Cotter quote haulage charges as: Kangu-Beretemba \$17.00; Kangu-Siwai \$10.00; Kangu-Buin \$1.50 (Couper and Cotter, 1972:68).

Taking the more favourable options on the costs of shipping cargo via Kangu, estimates are obtained as shown in Table 7.9.

In 1972 one Kieta-based road haulage company was regularly carrying freight between Kieta and Southwest Bougainville. The two northern co-operatives (at Beretemba and Konga) were using this means of shipping cocoa to Kieta. The road freight charges were \$12.00 per ton to or from Beretemba or Konga, and \$17.00 per ton to or from Buin. Assuming that insurance was the same as for transport by sea, and that unloading charges into store at Kieta were included in the freight rate, the costs of shipping by road are as given in Table 7.10. Hence it is significantly chaper to use road transport (at the prevailing rates) from the northern two-thirds of the area, and only marginally more expensive from Buin. The

Table 7.9

Estimated costs of shipping by sea,
Southwest Bougainville-Kieta
$/ton

Elements of costs	From Beretemba	From Konga (Siwai)	From Buin
Road haulage to Kangu	14.00	6.00	1.50
Stevedoring at Kangu	1.60	1.60	1.60
Sea freight to Kieta	11.50	11.50	11.50
Wharfage and handling, Kieta	1.80	1.80	1.80
Cartage ship to shore, Kieta	0.20	0.20	0.20
Insurance	3.00	3.00	3.00
Totals	32.10	24.00*	19.60

* This figure compares with Couper and Cotter's figure (1972: 70) of $25.43 for total transport costs on cocoa from Siwai to Kieta via Kangu, although their bases are slightly different.

Table 7.10

Estimated costs of shipping cocoa by road,
Southwest Bougainville to Kieta
$/ton

Elements of costs	From Beretemba	From Konga	From Buin
Road haulage to Kieta	12.00	12.00	17.00
Insurance	3.00	3.00	3.00
Totals/ton	15.00	15.00	20.00
Costs/ton mile	33 cents	23 cents	22 cents
Difference from sea route	-$17.10/ton	-$9.00/ton	+0.40c/ton

indications are, therefore, that the northern two-thirds of the area, i.e. Nagovisi and Siwai, will switch immediately to road transport, but the position for the southern third is somewhat more delicate.

All of the area, if it switches to road transport, is at risk because of the closing of the trans-island road by bad weather, either through rivers flooding or by slips on the hill section near Sikoreva. However, north of Siwai there is only one major unbridged river crossing, that of the Pangala River, west of Koro village. The permanent bridging of this river and the improvement of the Sikoreva hill section should therefore have a fairly high priority in assessing construction needs of the Southwest. An obvious preliminary step will be to keep a check on the delays caused by flooding or weather conditions on these sections. South of Konga, however, there are at least four major rivers likely to disrupt traffic by flooding, and the only solution in the long run is the construction of a new Konga-Buin road some miles inland. This is likely to be a more costly undertaking than bridging the Pangala and hence likely to be longer in eventuating.

Couper and Cotter recommend that a shipping service by a landing barge type vessel of about 50 NRT should be maintained to Kangu in the short term to meet the inward and outward cargo needs of the southern third of Southwest Bougainville which are at present about three-fifths of the whole. Given the fact that flooding does occur frequently on the four southern rivers[1] this recommendation is reasonable for the short term, but the poor quality of Kangu as a port together with the likelihood of increased cocoa production throughout the Southwest indicates that a new inland road between Konga and Buin should be built in the longer term.

[1] In May 1972 the rivers made the road impassable for about four consecutive weeks. A reliable official record should be maintained of the number of days the road is impassable.

Chapter 8

Benefit-cost studies on trunk and feeder roads

The quantification of benefits derived from building new roads or upgrading existing ones is difficult due to the subjectivity involved in deciding what benefits are directly attributable to the project in question and what values should be placed on them. Nevertheless, the technique of benefit-cost analysis is widely used to assess the returns from construction projects. Benefit-cost analysis can be extended to give internal rates of return and from these an order of priority for a series of projects can be determined in terms of relative benefits. Sensitivity analyses can also be performed on the costs and benefits of a project to show what variations may be expected in the internal rate of return under changing values of costs and benefits.

A number of benefit-cost analyses have been carried out for elements in the road system of Southwest Bougainville. In the trunk road system these included the existing trans-island road; a possible higher-quality trans-island road over an improved route in the same locality; and the proposed Buin-Toimonapu road across Kono census division in the south-east of the island.

In addition benefit-cost studies were carried out on feeder road networks in the Nagovisi-Baitsi, Siwai and Buin areas, since the improvement of feeder roads is an important pre-requisite to increasing cocoa production in the area and to the well-being of the inhabitants.

Benefit-cost analysis of the existing trans-island road

The road under consideration was built in 1972 linking Moratona with the Jaba River Pump Station. Its length is 14 miles.

Benefits. Only economic benefits were considered, namely the savings in transport costs on cocoa and copra outwards and on general cargo inwards, as compared with the only previous method of shipment - sea transport from Kangu and use of the internal road system. Because the economic benefits are very large it was not necessary to consider social benefits in this case.

Assumptions. The following assumptions were made:

(1) That all production from the Bana and Siwai areas is now using this road.

(2) Projections of cocoa production were calculated on the basis of new plantings 1966-69 reaching maturity in five years, and the same rate of growth projected by trend-line analysis to 1982 (Table AV.1).

(3) Copra production was assumed to double its 1971-72 volume by 1975-76 and then remain steady (Table AV.2).

(4) Freight savings were assumed to be $17.10/ton from Beretemba and $9.00/ton from Konga to Kieta (Tables 7.9 and 7.10). A lower insurance rate for cargo travelling by road would increase the savings and hence increase the benefits, but this has not been included.

(5) That one-third (1,184 tons) of total imports through Kangu (3,551 tons in 1972) are destined for the Bana area and one-third for the Siwai area.

(6) The growth rate of imports through Kangu 1969-70 to 1971-72 was 15 per cent per annum. A future growth rate of 7.5 per cent per annum was assumed for the first five years and 4 per cent per annum thereafter.

(7) That the same savings in transport costs for imports as for the outward shipments apply. Hence savings start at $22,614 for the Bana area and $10,656 for Siwai.

(8) A life of 10 years for the road was assumed, and its residual value was ignored.

Costs. Although there is some uncertainty about the figure, a cost of $60,000 was assumed for the construction of the road. Annual maintenance costs were assessed to be $1,000/mile.

Table AV.2 gives the calculation of economic benefits (freight savings) for the existing trans-island road. Table AV.3 gives the calculation of total economic benefits, net benefits, and present worth of net benefits at a 20 per cent discount rate.

Benefit-cost ratio (BCR). Hence the benefit-cost ratio at 20 per cent discount rate is 198,607:60,000 = 3.31.

Internal rate of return. The internal rate of return is 67 per cent.

The internal rate of return is high, but this is expected because the opening of the 1972 trans-island road was a major improvement to the existing transport system. It therefore captured a large proportion of the benefits that were nascent in the existing situation. Because it now exists, any subsequent major improvement will be able to attract only lower benefits as long as the existing road remains open. If the present trans-island road were to deteriorate so that it became negotiable only with great difficulty or not at all, then alternative road routes would attract higher benefits than they do at present.

Because the present trans-island road is of low quality and is likely to carry relatively large volumes of traffic, a road of higher standard will ultimately have to be built to replace it. Thus a benefit-cost analysis has been carried out for a new road of higher quality.

Benefit-cost analysis of a new trans-island road

The road under consideration would be of Rural II standard, including crossings of the Pangala and other rivers, and include minor upgrading of the road between the Jaba River Pump Station and Panguna. A new route is assumed, the location of which would need to be determined by engineering survey, but which would lie roughly between Beretemba-Sovele-Waitabuna-Momogonari-Jaba Pump Station.

It is assumed that 10 miles of new road and 5 miles of upgrading (between the Pump Station and Panguna) would be necessary.

Benefits. The benefits derive from savings on transport costs on cocoa and copra outwards and general cargo inwards, compared with the existing trans-island road; and from social travel benefits.

The economic and social travel benefits arise because the proposed new road would reduce the distance between Beretemba and the Jaba River Pump Station from 23 to 10 miles (over the existing trans-island road) and the section from the Pump Station to Panguna from 6 to 5 miles; and because the road at Rural II standard reduces vehicle operating costs.

Assumptions. (1) Freight rates: In 1972 actual road haulage charges to Kieta by the existing road (omitting insurance) were:

From Beretemba	\$12/ton
From Konga	\$12/ton
From Buin	\$17/ton

Since Beretemba to Konga is a distance of 19 miles the actual cost from Konga to Kieta should be more than from Beretemba to Kieta, and it is arbitrarily assumed to be \$14/ton.

The distances by the 1972 road are:

Beretemba-Kieta	46 miles
Konga-Kieta	65 miles
Buin-Kieta	90 miles.

Hence the average ton mile rates by the existing road are:

Beretemba-Kieta	26 cents/ton mile
Konga-Kieta	22 cents/ton mile
Buin-Kieta	19 cents/ton mile.

The distances are reduced by 14 miles by the new road.

It is further assumed that the ton mile rates on the 15 miles of new and upgraded road between Beretemba and Panguna are reduced to 12 cents/ton mile.

Thus the savings on the transport of freight are:

From Beretemba: (29 miles x 26 cents/ton mile) − (15 miles x 12 cents/ton mile) = \$5.74/ton

From Konga: (29 miles x 22 cents/ton mile) − (15 miles x 12 cents/ton mile) = \$4.58/ton

From Buin: (29 miles x 19 cents/ton mile) − (15 miles x 12 cents/ton mile) = \$3.71/ton

It is assumed that a difference of this magnitude compared with sea rates from Buin to Kieta is sufficient to induce Buin cargoes to move by road.

(2) Tonnages: The production of cocoa in Bana, Siwai and Buin areas is projected on the basis of known new plantings for 1966-69 for Bana and Siwai, and 1971-73 for Buin, with continued growth assumed. (The growth rate is about 6 per cent per annum.)

The production of copra in all three areas is assumed to double its 1972 value by 1976 and then remain constant. (High prices for copra in 1973-74 may induced increased production in Southwest Bougainville and hence increase the benefits.)

Inwards freight is assumed to be one-third of Kangu imports in 1972 to each of Bana, Siwai and Buin areas. Future growth is assumed to be 7.5 per cent per annum from 1972-77 and 4 per cent per annum thereafter. The same savings in transport costs for imports as for outward shipments are assumed to apply.

Table AV.4 calculates total tonnages, freight savings and economic benefits for the new road.

(3) Social travel benefits: 'Social' travel benefits have been calculated on the basis of vehicle operating cost savings on journeys made for purposes other than the carriage of cocoa or copra outwards or general cargo inwards. Some 'social' journeys will be made by people travelling as passengers on cargo carrying trips, but it is impossible to evaluate the number, and there are no directly comparable data to apply from other studies. Accordingly, total social travel benefits have been lowered by 10 per cent to allow for this.

The destination for 'social' journeys is assumed to be a point halfway between Panguna and Kieta, on the assumption that half the passenger journeys will be to Panguna and half to Kieta. Hence the distances assumed for social travel are:

From Beretemba	28 miles
From Konga	47 miles
From Buin	72 miles.

Social journeys to visit Panguna, Arawa or Kieta for such purposes as shopping, medical treatment, education, government business, etc., may be expected to start from a low base, but increase over the 10-year period as the area develops and people become more mobile. About two years after the Rigo road was opened for traffic it was found that the number of return journeys made to Port Moresby from the area served by the road was 1 per week for every 12 people (Ward, 1970:73).

It is therefore assumed conservatively that the people of Southwest Bougainville will use the trans-island road at one-quarter this rate, namely 1 journey per week for every 48 people, or 50 journeys per year for every 48 people. But it is also assumed that this rate is not reached for nine years. Hence the number of 'social' journeys is taken as:

10 journeys/year for every 48 people in the 1st year
15 " " " " " " " " 2nd year
20 " " " " " " " " 3rd year
25 " " " " " " " " 4th year
30 " " " " " " " " 5th year
35 " " " " " " " " 6th year
40 " " " " " " " " 7th year
45 " " " " " " " " 8th year
50 " " " " " " " " 9th year

The total number of people concerned (reckoning population growth at 4 per cent per annum) is:

	1972	1977
Bana area (omitting Northern Banoni)	8,346	10,154
Siwai	8,559	10,413
Buin area (half the total, to allow for greater distance, hence lower likelihood of visits)	6,513	7,924

These totals are assumed to grow at 4 per cent per annum.

For the calculation of social travel benefits vehicle operating cost (VOC) savings are taken from tables in the UNDP report (Halcrow, 1969, vol.2:7.53). The relevant sections are:

VOC for utilities on 'clay surface, steep, winding, bad alignment'	= 43.8 cents/mile
VOC for utilities on 'good, gravel surface, hilly, curved, good alignment'	= 22.5 cents/mile

Hence VOC savings are 21.3 cents/mile. However, a conservative figure of three-quarters of this value, i.e. 17.5 cents/mile, is taken in the present calculations.

The average load factor for passengers is assumed to be 2.5 per vehicle.

Hence social travel benefits start at the following levels and rise as described above, with an increase in population of 4 per cent per annum.

For Beretemba $$\$\ \frac{17.5 \times 28 \times 2 \times 10{,}154 \times 10}{100 \times 48 \times 2.5} = \$8{,}296$$

For Konga $$\$\ \frac{17.5 \times 47 \times 2 \times 10{,}413 \times 10}{100 \times 48 \times 2.5} = \$14{,}276$$

For Buin $$\$\ \frac{17.5 \times 72 \times 2 \times 7{,}924 \times 10}{100 \times 48 \times 2.5} = \$16{,}640$$

The calculation of social travel benefits for the new trans-island road is given in Table AV.5.

Costs. (a) Construction: The costs for construction of a Rural II standard road between Beretemba and the Jaba River Pump Station, including a crossing of the Pangala River, are estimated following Smith (1972:25-7). They are based on the actual costs of building the Konga-Boku road in 1968.

	Rates/mile Barclays 1968 contract $	1968 contract plus 5% p.a. escalation over 7 years $
New construction to Rural II standard		
Roadworks and culverts	33,000	47,000
Bridges and causeways	17,000	24,000
	50,000	71,000
Cost estimate for 1975 construction: 10 miles at $71,000/mile		$710,000
Bridges and approaches: Rate/foot of main span	1,000	1,500
Causeways and approaches: Rate/foot of actual crossing	100	140
	Bridges $	Causeways $
River crossings necessary:		
Munara 40' assumed bridge span) Konara 40' " " ")155' Orei 75' " " ")	232,500	24,500
and either Pangala 180') and Nunopa 140') 320'	480,000	44,800
or Pangala 250'	375,000	35,000

Costs	either	or
Bridges	$712,500	$607,500
Causeways	$69,300	$59,500

Hence costs for the new section range from:

Minimum	$710,000 + 59,500 = $769,500
Maximum	$710,000 + 712,500 = $1,422,500

An allowance of $100,000 is made for upgrading the Jaba River Pump Station to Panguna sections. Total estimated costs thus range from $869,500 to $1,522,500. Construction costs are assumed to be spread over the three years 1974-77 in the ratio 1:2:1. The discounted values are given in Table AV.6.

(b) Maintenance. Maintenance costs are based on the UNDP formula (Halcrow, 1969:3, 13.26 and 13.33).

Annual maintenance costs/mile = \$(400A + B.Q)

where A = coefficient for terrain (in this case, 'stable-hilly', A = 2)
B = coefficient for road standard (in this case 'new secondary', B = 5 to 6, say 5.5)
Q = average daily traffic, unweighted all vehicles (in this case, say 50 u.p.d.)

Hence maintenance costs per mile = \$(400.2 + 5.5.50) = \$1,075. Over 15 miles this amounts to \$16,125 per annum.

Table AV.7 gives the calculation of total benefits, net benefits and present worth of new benefits at discount rates of 5, 10, 15 and 20 per cent.

Benefit-cost ratio and internal rate of return. The benefit-cost ratios and internal rates of return for the maximum and minimum costs at various discount rates are given below.

	Discount rate					IRR
	0%	5%	10%	15%	20%	
Maximum cost benefit-cost ratio	1.12	0.77	0.55	0.40	0.30	1.5
Minimum cost benefit-cost ratio	1.76	1.35	0.96	0.70	0.53	9.0

Thus the internal rates of return are relatively low and indicate that only a minimum cost road should be considered. As was pointed out above, the low figures result from the presence of the existing road as well as from the conservative nature of the assumptions. The high maximum cost derives mostly from the inclusion of provision for bridges rather than causeways across the major rivers. The decision whether or not to build bridges rather than causeways should be an engineering one, but in view of the importance of the trans-island road to the whole trunk road system of Bougainville Island it seems that major rivers (particularly the Pangala) should be crossed by bridges so that the road can remain open under all weather conditions.

Sensitivity analysis. A sensitivity analysis has been performed for the minimum cost case of this road using ratios of actual/estimated cost of 1.0, 0.9, 0.8 and 0.7 and of actual/estimated benefits of 1.2, 1.0, 0.8 and 0.6. The details are given in Appendix V and Fig. AV.1.

Benefit-cost analysis of the Buin-Toimonapu road

A benefit-cost analysis has been carried out on a possible road link between Buin and Toimonapu in the southeast of the island. The precise route would depend on engineering surveys but would be approximately from Buin town through Tabago and Ububakogu villages to Toimonapu, the road head on the east coast about 25 miles (40 km) south of Kieta. This road would be a link in the round island trunk road which will probably be achieved in the long term. The length of new road construction necessary would be about 23 miles (37 km). The two major physical difficulties are a crossing of the Deuro Range at about 600 metres and a crossing of the Luluai River. Reconstruction of the existing road between Toimonapu and Aropa airfield would also be required. This is omitted from consideration here because benefits would accrue from this to villages and plantations along it, for which data have not been collected.

For calculation purposes, it is assumed the road is built between 1975 and 1978, with expenditure in the proportions 1:2:1 over those years.

Benefits. Benefits would accrue from the savings on transport costs of cocoa and copra production in the Makis, Lugakei, Paubakei and Kono census divisions, and of general cargo inwards to those areas, compared with present transport costs by sea through Kangu, and from social travel.

Assumptions. (1) Cocoa production is assumed to increase at a rate based on 1965-69 and 1971-73 new plantings in the Buin area reaching maturity in five years, projected by trend-line analysis. The projected volumes of cocoa production are given in Table AV.8.

(2) Copra production is assumed to double by 1975-76 and then remain steady.

(3) Inwards freight is assumed to be one-third of the amount entering Kangu in 1972, projected at 7.5 per cent per annum to 1977 and thereafter at 4 per cent per annum.

(4) Rates on the proposed road for both inwards and outwards freight are assumed to be 15 cents/ton mile. The distance from Buin to Kieta via the proposed road is 23 + 25 = 48 miles. Hence the freight cost per ton from Buin to Kieta would be $7.20. Sea freight (less insurance) from Kangu to Kieta was $16.60 in 1972. Hence freight savings would be $9.40/ton.

(5) Social travel benefits are calculated in the same manner as for the new trans-island road. The 1972 Buin population was

13,026, which would be 16,482 by 1978-79. A population growth rate of 4 per cent per annum is assumed. Hence social travel benefits start at

$$\$ \frac{17.5 \times 48 \times 2 \times 16{,}482 \times 10}{100 \times 48 \times 2.5} = \$23{,}075$$

Tables AV.9 and AV.10 calculate economic and social travel benefits for the Buin-Toimonapu road.

Costs. Construction costs are calculated as for the new trans-island road:

Roadworks: 23 miles at $71,000/mile = $1,633,000.
Bridges and causeways: the necessary river crossings are:

River	Assumed length (feet)	
Muliko	50)	
Samiei	50)	
Aobai	50)	
Atara	50)	500
Abia	50)	
Pirias	50)	
Luluai	200)	

A figure of $1,500/foot is assumed for bridges and $140/foot for causeways.

Hence costs, if causeways: 500 x 140 = $70,000
if bridges: 500 x 1,500 = $750,000.

Hence total cost range is:

Minimum: $1,633,000 + 70,000 = $1,703,000
Maximum: $1,633,000 + 750,000 = $2,383,000

The calculation of the present worth at 5 per cent discount rate of the minimum cost estimate is given in Table AV.11. Maintenance costs, calculated at $1,075/mile, are $24,725 per annum.

The calculation of total benefits, net benefits and present worth of net benefits at discount rate of 5 per cent, is given in Table AV.12.

The benefit cost ratios for the minimum cost estimate are

at 0% BCR = 0.49
at 5% BCR = 0.38

Hence the road is clearly uneconomic on calculations based only on cost savings for the southern third of Southwest Bougainville as compared with sea transport. As pointed out above this is due to the fact that the existing trans-island road has tapped off the potential traffic, and hence benefits, by virtue of its prior construction. Nevertheless, a Buin-Toimonapu road does have merit as a necessary link in the completion of a round-island trunk road system.

It also has an important role in providing road access for the 1,166 villagers (in 1972) of the isolated Oria area. The benefits, which would derive from the transport of cocoa (estimated at 75 tons production per annum by 1977), plus the value of new cocoa production generated in this area by road access, plus the freight savings on the outward transport of fresh fruit and vegetables presently being flown out of this area, and the potential increase in this activity, plus savings on inwards freight, have not been included in this calculation, since they are more correctly attributed to the feeder road construction which would be required to link the villages with the main Buin-Toimonapu road. These benefits, however, are not large enough to make the Buin-Toimonapu road an economic proposition at the present time.

If a feeder road is built to link the Oria area with Buin it is important that it should be located on the route selected by engineering appraisal for the Buin-Toimonapu road. The completion of that link would then become an attractive proposition.

Benefit-cost analyses for other trunk roads

Benefit-cost analyses have been carried out for two other sections of the trunk road network in Southwest Bougainville, namely the Boku-Konga road constructed in 1968, and the Konga-Buin link as yet unconstructed. In both cases they resulted in low benefit-cost ratios.

On economic considerations, the Boku-Konga link was clearly constructed prematurely, since the savings gained were only those in vehicle operating costs over a relatively small reduction in length (to about 16 miles (26 km) compared with the previous 21 miles (34 km)) and improvement in quality, including the elimination of about four major river crossings. These savings were felt mainly on traffic to and from the Bana area, although the Siwai census division has benefited by gaining an all-weather road running through the heart of its most densely populated area. Consequently improvement to all-weather reliability in the feeder road system serving the main road in Siwai now becomes an important goal.

The Konga-Buin trunk road also now shows low benefit-cost ratios since in the present system it can only attract benefits from savings in vehicle operating costs for traffic from the Buin sub-area. Again the distance savings are small, but the advantages to be gained by the elimination of the dangerous crossings of five major rivers (the Silibai, Porror, Oamai, Mivo and Mobiai) are very important. The proposed route runs about 6 km inland of the present road and hence nearer the centre of population in the Makis census division. Despite the low benefit-cost ratio this road could attract it is a necessary part of the trunk road system, and it will eventually have to be built. Hence it is important that feeder roads which may be built to link Buin town with Kanauro, Tugiogu and Turigau villages should follow the proposed route of the Konga-Buin trunk road. A benefit-cost appraisal has not been carried out for a section of trunk road leading north from say Moratona across Northern Banoni to Torakina and northwest Bougainville. This road would pass through lightly populated country for most of its length, and the local benefits it would attract would be small. Nevertheless such a road should be an ultimate aim in trunk road construction for Bougainville Island.

In summary the priorities for trunk road building in Southwest Bougainville at the present time appear to be:

(1) a new trans-island road
(2) the Konga-Buin link
(3) the Buin-Toimonapu link
(4) a trunk road from Moratona to the northwest.

It may well be the case that other sections of the round-island trunk road in northern Bougainville are more urgent than any of these, providing the existing trans-island road can be kept open.

Benefit-cost analyses of feeder roads systems

In view of the fact that Southwest Bougainville now has a road link to Kieta the improvement of existing and the construction of new feeder road systems becomes of higher priority than any further trunk road construction in this area in the immediate future. This is particularly so if the large existing resource of cocoa plantings is to be fully realised in the next decade.

Consequently feeder road systems have been devised for each of the three sub-areas and subjected to benefit-cost analysis. These systems, which are shown in Figs 8.1, 8.2 and 8.3, have

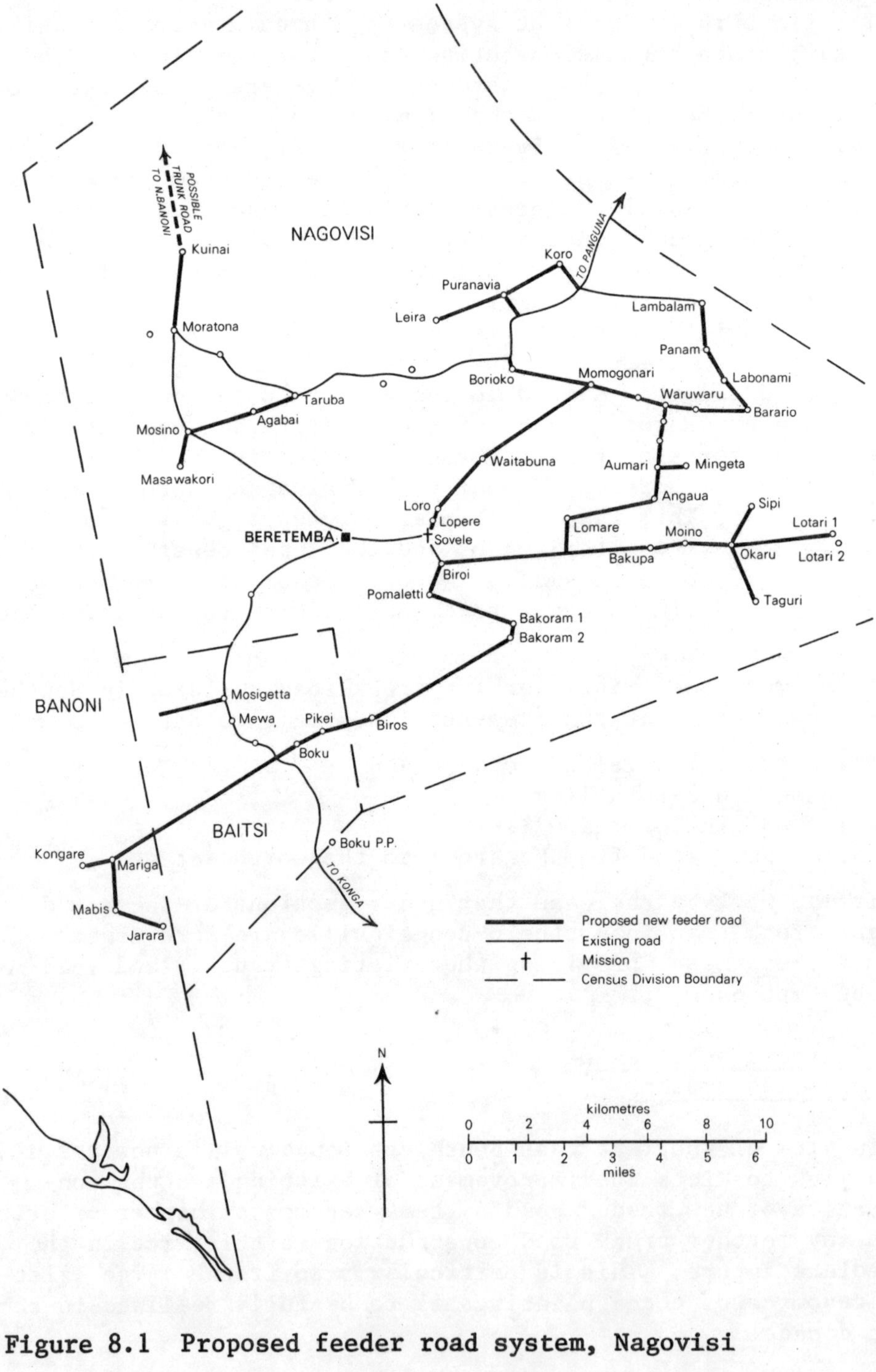

Figure 8.1 Proposed feeder road system, Nagovisi

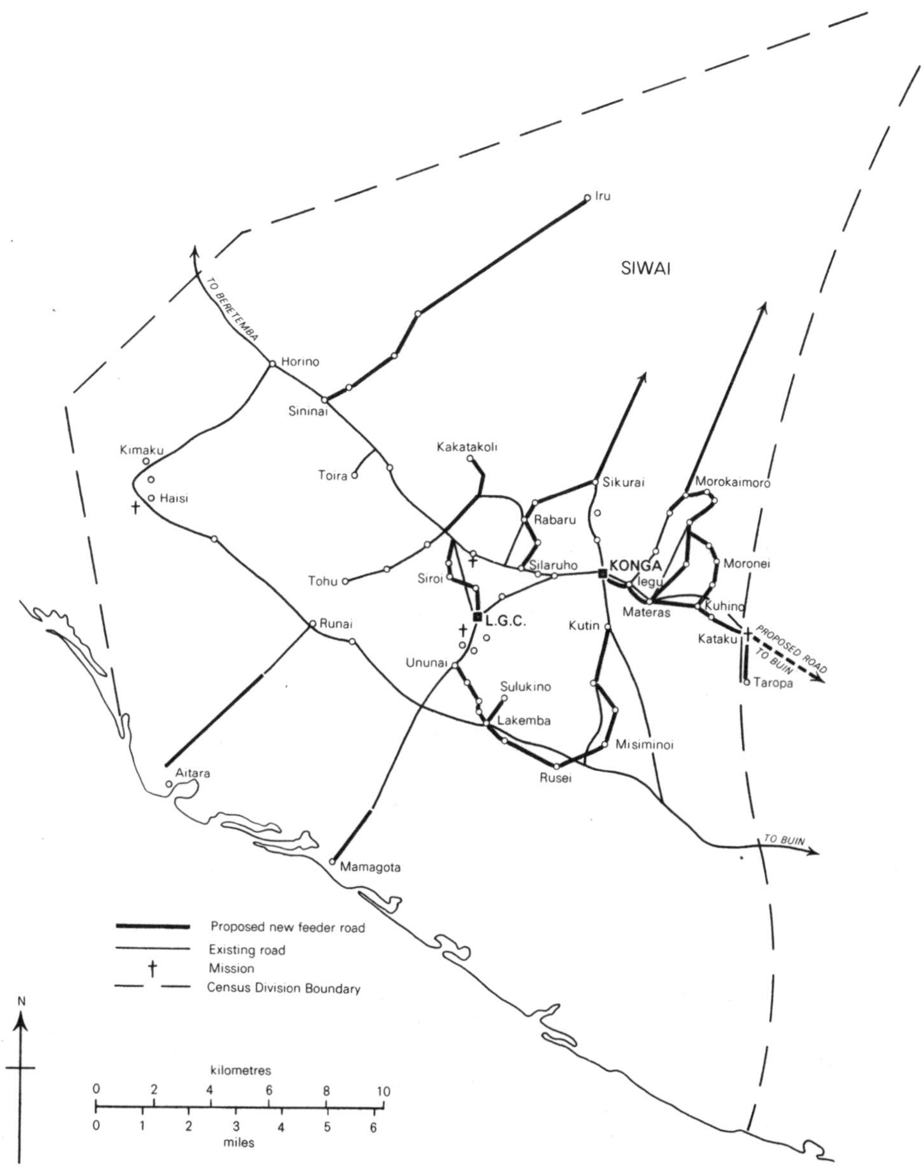

Figure 8.2 Proposed feeder road system, Siwai

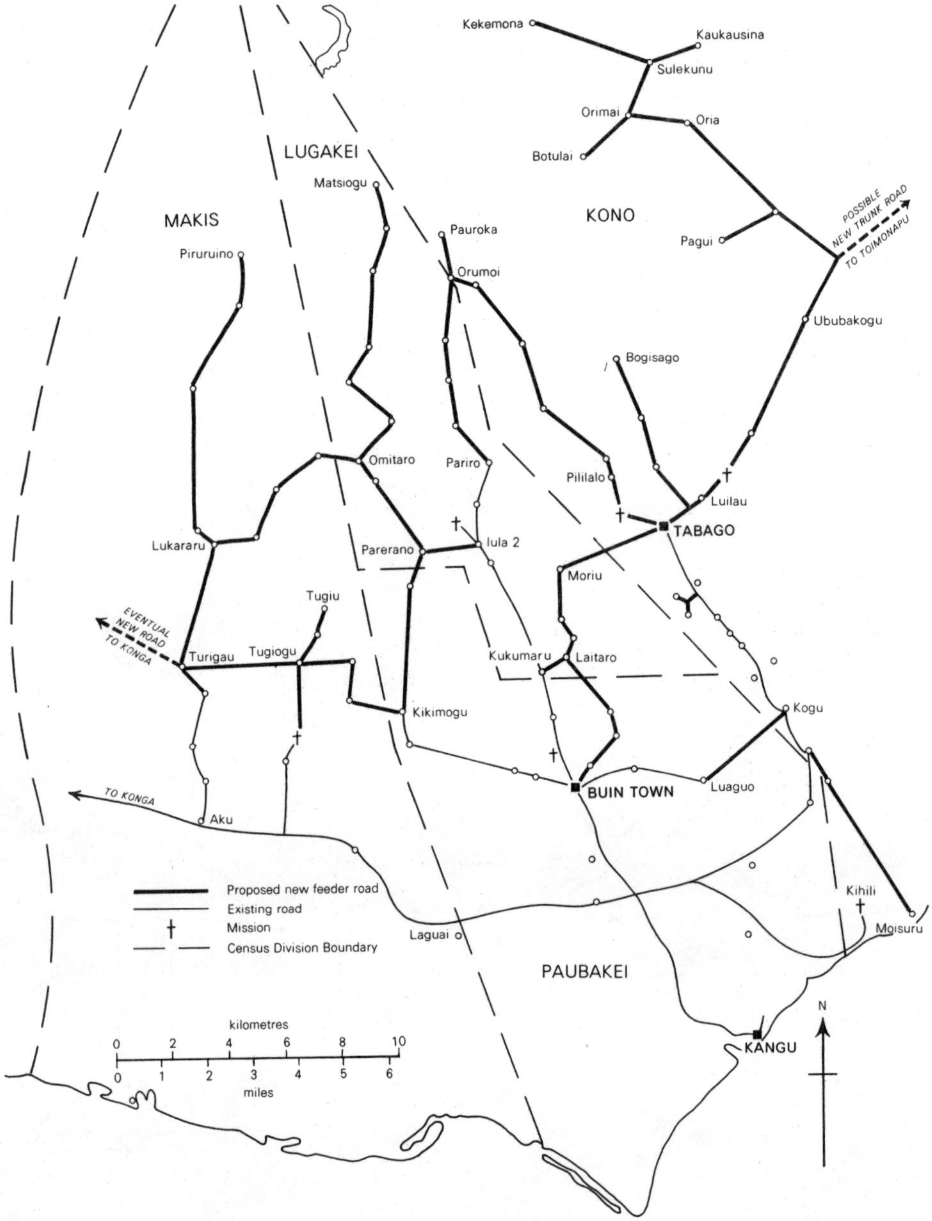

Figure 8.3 Proposed feeder road system, Buin sub-area

Plate 15. A mudhole on the feeder road near Iegu village east of Konga, Siwai census division

Plate 16. Local transport on a feeder road near Konga, under dry conditions.

been drawn up without benefit of engineering advice as to the best routes, and may consequently not represent the best precise ultimate locations of the roads. They have been designed on economic, social and locational grounds and are intended to provide where possible circuits rather than dead-end roads, focusing as much as possible on the incipient urban centres of each sub-area, namely Beretemba, Konga and Buin town. In some places feeder roads already exist, but except in some areas of easy country in Siwai census division they are mostly in need of upgrading to bring them to near all-weather negotiability. Some improvement work has been carried out recently and is continuing. Existing feeder roads have been incorporated in the new systems proposed.

Information available. The only information available on which to base benefit-cost analyses of feeder road systems was:

(1) The number of cocoa trees by village, for 1970 in the cases of Bana and Siwai and for 1973 for the Buin sub-area. These counts are approximate, but nevertheless are probably reasonable indicators of the amount of cocoa planted.

(2) Village populations in 1972, and a probable growth rate of 4 per cent per annum.

(3) The approximate miles of road needed in each case.

(4) Approximate costs, which have been taken at \$5,000/mile, whether for new construction or upgrading existing road; and maintenance at \$300/mile per annum.

It was therefore necessary to devise a formula utilising these known data, which could be applied over the whole set of roads.

Benefits have been assumed to accrue from:

(1) Vehicle operating cost savings in carrying wet bean to the central fermentaries at Beretemba, Konga and Buin town.

(2) Value added of generated production, assumed to be 30 per cent above the present figure.

(3) Vehicle operating cost savings on 'social' travel.

Assumptions. (1) Production of wet bean is assumed to be 6 lb/tree per annum, so that tons of wet bean produced =

$$\frac{\text{no. of trees} \times 6}{2{,}240} = \frac{\text{no. of trees}}{373}$$

(2) The amount of wet bean carried to a fermentary is assumed to be 0.5 ton per vehicle trip. Utilities are capable

of carrying more but cocoa production occurs throughout the year (ignoring the flush periods for the present purpose). The number of vehicle trips necessary to carry wet bean from the villages to the fermentaries is thus 2 (tons cocoa).

(3) It is assumed that each vehicle trip is a return trip, so that the distance covered on each trip = twice the mileage of each road or circuit touching all villages and the fermentary.

(4) The vehicle operating cost savings are assumed to be (where a road already exists) the difference between the vehicle operating cost for utilities on clay surface, steep, winding, bad alighnment = 43.8 cents/mile, and gravel surface, hilly, curved, good alignment = 24.9 cents/mile (Halcrow, 1969:vol.2:7.53), i.e. 18.9 cents/mile. Where no road exists half this value is taken, i.e. 9.5 cents/mile.

(5) An allowance of 30 per cent of existing production is made for the new production generated by the presence of a good feeder road. This is valued at $44.80 per ton of wet bean, i.e. at the farm-gate price of 0.5 cents/lb.

(6) Road construction is assumed to be completed in the one year of 1975, and hence present worth of benefits has been calculated back to that year.

(7) Cocoa production is assumed to increase at 6 per cent per annum both between 1970 and 1975 and for the 10-year period 1976-85, which is assumed as the life of the road. Residual values for the roads are ignored.

Hence the economic benefits are calculated to start at the following level, and increase at 6 per cent per annum:

$$\$2 \times \frac{\text{no. of trees}}{373} \times 2 \times \text{mileage} \times \text{either } \frac{18.9 \text{ or } 9.5}{100}$$

$$+ \frac{30}{100} \text{ of } \frac{\text{no. of trees}}{373} \times 44.80$$

'Social travel' benefits

Assumptions. (1) The number of journeys made for 'social' purposes is assumed to be 1 return journey per week for every 25 people in the area served by the road. This may be a slightly high estimate and social travel benefits could well have been reduced by 10 per cent for each year as was done for the trunk road sections. This would result in rather lower social travel benefits, but would not change the order of priority over the whole set of feeder roads.

(2) The VOC savings are assumed as in (4) above to be 18.9 cents/mile if a prior road exists and 9.5 cents/mile if no road exists.

(3) 'Social' travel is assumed to grow at the rate of population growth, i.e. 4 per cent per annum.

Hence social travel benefits are calculated to start at the following level and grow at 4 per cent per unit per annum.

$$\frac{\text{No. of people}}{25} \times 50 \times \frac{18.9 \text{ or } 9.5}{100} \times 2 + \text{no. of miles}$$

The calculations of benefit-cost ratios for the whole series of feeder roads are given in Appendix V, Schedules 1-22.

The internal rates of return (IRR) for each road are given in Table 8.1. These internal rates of return provide a priority ranking for construction or upgrading in terms of the economic and social benefits to be gained from each road. Section A of the table lists those roads with benefit-cost ratios greater than 1, and Section B those with benefit-cost ratios less than 1 at 5 per cent discount rate.

The benefit-cost analysis places emphasis on the economic benefits to be gained from the transport of cocoa from village to fermentary, and on social travel benefits gained from the movement of people for non-economic reasons. Hence the areas which have large plantings of cocoa and lack roads such as inland Nagovisi and the Buin sub-area show high economic benefits. On the other hand the Siwai area has a dense population although less cocoa, and hence the 'social travel' benefits exceed the economic. Since much of Siwai already has an existing low-quality feeder road network the strategy which would bring greatest immediate benefit to most people would be to concentrate on the roads in the Nagovisi and Buin sub-areas.

The layout of roads shown here has been deliberately designed to provide circuits rather than dead-end roads wherever possible. The intention is to provide greater opportunities for circulation of people and services amongst villages. In the steeper inland areas it is likely to be impossible to complete circuits since the steep-sided ravines in the terrain would be expensive to cross by roads. In Siwai two dead-end roads have been shown leading to inland areas since these are areas of potentially good land, which may need to be utilised in the future.

The feeder road system proposed here would provide road access to almost all villages in Southwest Bougainville, with

Table 8.1

Internal rates of return

Priority	Sub-area	Road	Length (miles)	Estimated capital cost	IRR (%)
A. 1	Siwai	Konga-Kutin-Kokui-Kunu-Misiminoi-Rusei-Moibinei-Lakemba-Sulukino-Harinai-Harikagu-Urunai-Hire-Koropo-Toitoi-Monoitu RCM-Amio-Siroi-Osokori-main road	12 (mostly existing)	$60,000	61
2	Buin	Buin town-Mamaromino (2)-Mamaromino (3)-Luaguo-Kogu-Muguai-Taguruai-Iamaru-Lakoei-Kikibatsiogu-Komai-Ka'amoro-Tubaro-Tabago-Moriu-Ipilai-Siul-Laitaro-Okoiragu-Kukumai-Mamaromino (1)-Buin town	20 (of which 12 new)	$100,000	41
3	Nagovisi-Baitsi	Biroi-Lomari-Angaua-Aumari-Mingeta-Nukui-Osiangi-Waruwaru-Berereke i-Monogonari-Waitabuna-Loro-Lopere-Sovele	14.5 (of which 2 shared with No.4. So say 13.5)	$67,500	35
4	Nagovisi-Baitsi	Momogonari-Berereke i-Waruwaru-Sianeki-Barario-Labonami-Panam-Lambalam-Sikoreva-Koro-Puranavia-Leira-Borioko-Momogonari	18 (of which 2 shared with No.3. So say 16)	$80,000	34
5	Nagovisi-Baitsi	Boku-Pikei-Biros-Bakoram (1) and (2)-Pomaletti-Biroi-Sovele	8.5	$42,500	33
6	Siwai	Konga-Iegu-Materas-Kuhino-Hino-Taropa; Laku-Moronei-Maintoita-Moisua-Hanong-(Materas)-Naronai-Morokaimoro-Mokolino-Rokuse-Turugum. Plus 4 miles inland.	18 (of which 4 new)	$90,000	31
7	Buin	Tabago-Pililalo-Parero-Kogwiriru-Mogoroi-Ororoi-Orumoi-Pauroko-Koniguru-Kumirogu-Kugugai-Pariro-Iula (1)-Iula (2)	14.5 (or which 13 new)	$72,500	29
8	Buin	Kikimogu-Mouakei-Ibiro-Tugiogu-Toaragai-Tugiu-Turigau-Tsimbo-Lukararu-Morula-Kogisagano-Lobigou-Omitaro-Numakei-Parerono-Kikimogu	23 (new)	$115,000	20.5
9	Buin	Tabago-Luilau-Kaitu-Lukauko-Bogisago	4 (new)	$20,000	20.4
10	Nagovisi-Baitsi	Bakupa-Moino-Okaru-Sipi-Taguri-Lotari (1) and (2)	8	$40,000	18.5
11	Buin	Buin town-Maramoku-Moro-Kanauro-Kikimogu-Piarino-Iula (2)-Iula (3)-Kukumaru-Laitaro-Buin town	15.5 (of which 10 new)	$77,500	17.5
12	Siwai	Konga-Kinirui-Munu-Sikurai-Matukoli-Rabaru-Kupingku-Silaruho-Purikoli-Kakatakoli-main road. Plus 2 miles inland	10.5 (of which 4 new)	$52,500	17.0
13	Nagovisi-Baitsi	Mariga-Kongare-Mabis-Jarara	6.5	$32,500	16.0
14	Buin	Tabago-Luilau-Iruniok o-Turitai-Ububakogu-Oria-Pagui-Orimai-Botulai-Sulekunu-Kaukausina-Kekemona	20 (new)	$100,000	11.8
15	Nagovisi-Baitsi	(a) Taruba-Agabai-Mosino-Masawakori (b) Moratona-Kunai (c) Mosigetta-1.5 miles west	3) 1.5) 6 1.5)	$30,000	9.0
16	Siwai	Horino-Kimaku-Tokunoitu-Haisi-Hiru Hiru	6	$30,000	7.5
		Other possible roads having benefit-cost ratios less than 1 at 5 per cent discount rate			
B. 1	Buin	Kogu-Maika-Tandareki-Togurau Plantation-Moisuru BCR at 5% = 0.76	5	$25,000	
2	Buin	Omitaro-Borubi-Okomo-Pamaiuta-Buraburunna-Togulegu-Matsiogu BCR at 5% = 0.28	7.5	$37,500	
3	Buin	Lukararu-Tubaru-Tubobisou-Siuru-Piruruino BCR at 5% = 0.46	7	$35,000	
4	Siwai	Main road-Nukui-Usokoli-Tohu BCR at 5% = 0.71	2.5	$12,500	
5	Siwai	Ununai-Mamagota-Kotu-Runai-Aitara BCR at 5% = 0.34	13.5	$67,500	
6	Siwai	Sininai-Kumuki-Misiminoi-Miheru-Iru BCR at 5% = 0.75	7 (of which 3 new)	$35,000	

the exception of northern Banoni and a few coastal villages in the south and west. An estimate of the costs of building such a system is given in Table 8.2.

Table 8.2

Estimated costs for a comprehensive system of feeder roads in Southwest Bougainville
$

Census division	Group A	Group B
Nagovisi-Baitsi	292,500	-
Siwai	232,500	114,500
Makis-Lugakei-Paubakei-Kono	485,000	97,500
Total	1,010,000	212,000

Hence an expenditure of $1,000,000 on feeder roads in Southwest Bougainville over the next few years would provide an adequate and nearly comprehensive system, bringing real benefits to most of the people of the area. A further $200,000 would virtually complete the system, providing road access to all but the most remotely located villages. At say $200,000 per year over six years, this should not be beyond realisation in terms of the national program, when it is recognised that this is one of the most potentially productive areas of the country.

Chapter 9

Conclusion

This study has endeavoured to meet two aims: first, to present base-line data relating to the state of Southwest Bougainville about 1972 and, second, to examine the existing transport system and the immediate transport needs of the area. Particular attention has been given to the impact of the opening of the trans-island road, and to the assessment of a program of feeder-road construction which will be necessary if the full benefits of the trans-island road are to be realised.

The trans-island road, even in its present low-quality condition, has had an immediate effect in significantly lowering the transport costs of the outwards shipment of cocoa and copra and the inwards shipment of general cargoes. The people of Southwest Bougainville have gained not only through these lower transport costs but also in greater reliability of transport and reduced risk of damage and loss to their freight. The road has also made possible much easier and cheaper movement of people. Personal mobility is likely to increase considerably over the next few years, and hence also the circulation of knowledge about other places and people, and of ideas and innovations. This movement of people will be a two-way phenomenon, in that it is now much easier for residents of Southwest Bougainville to visit the major urban centres of Panguna, Arawa and Kieta, and conversely for residents of those towns to visit Southwest Bougainville. This may bring more money into the area but may also bring a certain degree of social disruption.

The opening of the road will have negative effects in that the use of Kangu as a port is likely to decline. The immediate prospect is that Kangu's trade will decline to one-third of what it was about 1972, and this may result in less frequent shipping calls there. External transport services for the Buin area may thus become worse than they were before the road was opened, and this may result in increased pressure from the Buin area for either the construction of a new road between Konga and Buin town, or for a new road to link Buin town with Toimonapu. The latter would greatly improve the external transport services for the isolated Oria area.

An overall effect of the opening of the trans-island road will be to shift the greatest volume of movement from the Buin area to the Nagovisi area (Figs 6.8 and 6.9). This will have the effect of encouraging the growth of Beretemba and to a lesser extent Konga/Amio, while possibly causing stagnation in the growth of Buin town.

A single development investment, such as the construction of the trans-island road, usually requires supplementary actions and investments to realise its full impact. In the present case the most urgent supplementary action appears to be the construction or improvement of the feeder road system. A set of priorities for feeder road building has therefore been developed. This places primary emphasis on building new roads into areas where there is much cocoa but poor access, particularly inland Nagovisi and middle and inland Buin. These roads can be strictly called 'development roads' in the sense that their construction is likely to result in increased production of cocoa and increased harvesting of the existing crop. Hence growers should experience a real improvement in their income. The particular circumstances of Southwest Bougainville may, however, require a solution to the apparent potential labour shortage if this goal is to be fully attained.

The pattern of feeder roads proposed here has been designed to stimulate a circulatory rather than a linear pattern of local movement. Consideration should be given to the establishment of mobile service units and _maket raun_ along the lines proposed for the Eastern Highlands (R.G. Ward _et al_., 1974). The centres of Buin town, Beretemba and Konga/Amio are suggested as bases for such units, the role of which would be to take services and social amenities to the people in the villages on a regular periodic schedule. Government and other services which would lend themselves to such periodic circulation include agricultural extension work, health services, co-operatives' marketing services, business development, postal services, mechanics to service village vehicles, entertainment, education and local village markets. The selection of suitable sites would require more detailed investigation, but a number of large villages at possible feeder road junctions readily suggest themselves.

Other needed supplementary actions to road building in Southwest Bougainville include further agricultural extension work for cocoa, copra, cattle and fruit and vegetable growing;

promotion of local business development particularly in the fields of transport and transport service industries, such as garages, petrol sales, mechanical repair and servicing and road-related service industries such as the provision of local government hostels, restaurants, cafes, and food-stalls. If the marketing of cocoa is to continue through co-operatives there will be a need for further training of staff to operate them competently.

Building new roads can only be one part of an overall program of area improvement. In the case of Southwest Bougainville they can do much to enable the people to capitalise on their existing soundly village-based cocoa producing economy.

Appendix I

Land systems

Coasts

Ja. 3. Jaba. Sandy, subparallel beach ridges up to 3 m high and 61-152 m apart; mid-height and tall forest (Cerbera-Colophyllum, Vitex-Pometia). Swampy swales; neutral peaty sands; mixed herbaceous vegetation (Leersia-Hanguana).

Swamps

Mo. 5. Moila. Permanent peaty swamps; areas of stagnant water with mixed herbaceous vegetation (Leersia-Hanguana, Lycopodium-Gleichenia) and tall grassland (Phragmites-Saccharum); areas of flowing water with tall forest (Terminalia brassii-Campnosperma).

Plains

Sa. 6. Saua. Flood-out plain of volcanic debris from braided streams below Mt Bagana; brown mottled sands; tall grassland (Phragmites-Saccharum). Recent ash-covered volcanic alluvial fan, grey fine sands; scrub (Cyathea-Bambusa).

Sl. 7. Silibai. Broad plain of volcanic-derived alluvium of low or high gradient with high water-table and subject to flooding; brown mottled sands; tall forest (mainly Vitex-Octomeles). Meandering rivers.

Si. 8. Siwai. Broad plain of ash-covered volcanic alluvium, of very high gradient, with very low water table; brown loams over sand; tall forest (Vitex-Pometia). Angular river channels.

Volcano-alluvial fans

Bu. 9. Buin. Partly dissected plains of very high gradient rising to plateaux of gentle slope; brown loams with an ash-pan; tall forest (Vitex-Pometia, Neonauclea-Sloanea) and secondary forest (Artocarpus-Albizia).

Nu.11. Numa. Long, low radiating ridges with very short, moderate side slopes; brown loams with lapillitic horizons; tall forest (Vitex-Pometia) and secondary forest (Artocarpus-Albizia).

Lk.12. Leikaia. Long, moderately high radiating ridges with short, very steep side slopes: brown loams with lapillitic horizons; tall forest (Vitex-Pometia, Neonauclea-Sloanea), and secondary forest (Artocarpus-Albizia).

Pr.13. Pauroka. Long, moderately high radiating ridges with steep to precipitous side slopes; brown loams; tall forest (Vitex-Pometia, Neonauclea-Sloanea), mid-height forest (Garcinia-Elaeocarpus), and secondary forest (Artocarpus-Albizia).

Volcanic landscapes

Ba.14. Balbi. Active or very recently active volcanoes; boulders or grey fine sands; bare, with mixed herbaceous vegetation, or with scrub.

Tk.15. Takuan. Extinct or dormant partially eroded volcanoes; brown loam; palm and pandan vegetation (Gulubia-Pandanus), mid-height forest (Garcinia-Elaeocarpus) and tall forest (Neonauclea-Sloanea).

Er.16. Erava. Moderately steep laharic fans largely dissected to ridges with precipitous slopes; brown loams; palm and pandan vegetation (Gulubia-Pandanus) and scrub (Cyathea-Bambusa).

Pu.18. Puto. Low or moderately high ridges of lava with steep slopes; brown loams over reddish friable clays or red friable clays; tall forest (Vitex-Pometia) and secondary forest (Artocarpus-Albizia).

Erosional hills

Ib.21. Ibu. Low dendritic ridges with both moderate and very steep slopes; brown loams with lapillitic horizons; tall forest (Vitex-Pometia) and secondary forest (Artocarpus-Albizia).

Mf.23. Mafahia. Moderately high radiating or dendritic ridges with very steep to precipitous short slopes; brown loams with lapillitic horizons; tall forest (Vitex-Pometia, Neonauclea-Sloanea).

De.25. Deuro. High dendritic ridges of coarse grain with steep slumping slopes; brown friable clays and brown loams over brown friable clays; tall forest (Vitex-Pometia).

Ba.26. Bagana. Moderately high dendritic ridges with very steep slopes subject to frequent ash showers; grey fine sands (stony phase); scrub (Cyathea-Bambusa).

Do.27. Doiabi. Moderately high dendritic ridges with steep upper slopes and very steep lower slopes; brown loams with lapillitic horizons; scrub (Cyathea-Bambusa).

Bo.28. Boira. High, narrow-crested ridges on older sedimentary rocks with very short very steep slopes; brown loams with lapillitic horizons; tall forest (Vitex-Pometia, Neonauclea-Sloanea) and secondary forest (Artocarpus-Albizia).

Mn.29. Mainoki. High, narrow-crested ridges on older sedimentary rocks with very short very steep slopes; brown loams or lithosols; tall forest (Vitex-Pometia) and secondary forest (Artocarpus-Albizia).

Erosional mountains

Me.33. Melilup. Very high, radiating or dendritic ridges with very steep upper slopes and precipitous lower slopes; lithosols; mid-height forest (Garcinia-Elaeocarpus), tall forest (Neonauclea-Sloanea) and scrub (Cyathea-Bambusa).

Ch.34. Chambers. High, radiating or dendritic ridges with very steep upper slopes and precipitous lower slopes; brown loams with lapillitic horizons; mid-height (Garcinia-Elaeocarpus) and tall forest (Neonauclea-Sloanea).

Ka.36. Karato. High dendritic ridges with very steep slopes; brown loams with lapillitic horizons; mid-height forest (Garcinia-Elaeocarpus) and secondary forest (Artocarpus-Albizia).

Pi.37. Pirurari. High dendritic ridges with very steep slopes; brown loams; mid-height forest (Garcinia-Elaeocarpus) and secondary forest (Artocarpus-Albizia).

Kt.39. Kieta. Dissected agglomerate plateau remnants with precipitous marginal scarps; brown friable clays; tall forest (Neonauclea-Sloanea), mid-height forest (Garcinia-Elaeocarpus); and secondary forest (Artocarpus-Albizia).

Karst plateau

Ke.40. Keriaka. Dissected limestone plateau with short very steep slopes and very high marginal scarps; brown loams with lapillitic horizons; scrub (*Cyathea-Bambusa*).

Appendix II

Village populations, 1972

Banoni

Village	Population	Village	Population
Kongare	126	Karekopa	124
Mabis	132	Mom	61
Mariga	82	Koiare	183
Jarara	188	Jaba	120
Mawareka	59	Piva	96
Matsunkei	84	Laruma	141
Tengerepaia	149	Kegiri	137
Sito	63	Beteriopaia	176
Korobi	123	Lesiopaia	144
Atangato	72	Keriana	154
Karatu	121		

Baitsi

Village	Population	Village	Population
Boku	147	Mosigetta	195
Lavoro	115	Pikei	257
Mewa	136		

Nagovisi

Village	Population	Village	Population
Agabai	109	Masawakori	92
Angaua	142	Mendai	143
Aumari	108	Mingeta	48
Bakoram 1	248	Moino	130
Bakoram 2	197	Mokokolio	140
Bakupa	160	Momogonari	101
Barario	144	Mosino	129
Berereke1	132	Nukui	180
Beretemba	228	Okaru	184
Biros	177	Osiangi	191
Biroi	105	Panam	177
Borioko	108	Pomaletti	202
Kuinai	91	Puranavia	118
Koro	195	Sianeki	141
Kupon	141	Sikoreva	115
Lambonami	153	Sipi	189
Lambalam	287	Sisiruai	113
Leira	141	Taguri	101
Lomari	135	Takimari	111
Lopere	136	Taruba	141
Loro	305	Waitabuna	124
Lotari 1	123	Waruwaru	252
Lotari 2	138		

Siwai

Village	Population	Village	Population
Aitara	95	Haisi	134
Amio	186	Hanong	93

Siwai (continued)

Hari	165	Moibinei	92
Harikagu	80	Mokolino 1	107
Hino	120	Mokolino 2	105
Hire	201	Morokaimoro	149
Hiru Hiru	146	Moronei	182
Horino	179	Munu	208
Hukoha	128	Musiminoi	198
Iegu	86	Naronai	73
Iru	152	Nukui	105
Kakatakoli	147	Osokori	81
Kapana	157	Panake	233
Kaparo	260	Purikoli	103
Kimaku	105	Rabaru	202
Kinirui	206	Rokuse	54
Kokui	157	Runai	71
Koropo	73	Rusei	308
Kotu	88	Sikurai	54
Kuhino	112	Silaruho	77
Kumuki	122	Sininai	98
Kunu	164	Siroi	192
Kupingku	83	Sulukino	137
Kutin	97	Tohu	67
Lakemba	116	Toira	66
Laku	120	Toitoi	124
Maintoita	83	Tokonoitu	139
Maisua	117	Tonu	118
Mamagota	101	Tonui	99
Materas	100	Turugum	192
Matukoli	218	Ununai	162
Miheru	128	Usokoli	131
Misiminoi	113		

Makis

Aku	170	Piruruino	61
Ibiro	73	Siuru	122
Kogisagano	89	Taropa	109
Kokopo	170	Toaragai	98
Laguai	446	Tokuaka	51
Lerot	123	Tsimbo	116
Lobigou	71	Tubaru	165
Lukaruru	257	Tubobisou	92
Morula	110	Tugiogu	311
Mouakei	83	Tugiu	260
Nabuku	185	Turigau	132
Nakorei	107		

Lugakei

Borubi	79	Matsiogu	106
Buraburunna	110	Moriu	72
Ipilai	190	Numakei	93
Iula 1	129	Okomo	94
Iula 2	117	Omitaro	57
Iula 3	83	Pamaiuta	111
Koniguru	198	Parerono	152
Kugugai	91	Pariro	378
Kukumaru	256	Siul	98
Kumirogu	76	Togulegu	37
Laitaro	271		

Paubakei

Kanauro	255	Maramoku	107
Kikimogu	148	Mituai	75
Kugiogu	91	Moro (Buin)	180
Kukumai	116	Nakano	113
Luaguo	124	Okoiragu	117
Malabita	229	Piarino	78
Mamaromino 1	188	Uguimo	43
Mamaromino 2	150		

Kono

Bogisago	228	Muguai	151
Botulai	106	Orava	153
Iamaru	83	Oria	157
Ka'amoro	196	Orimai	273
Kaitu	96	Ororoi	103
Kaukausina	83	Orumoi	73
Kekemona	102	Pagui	195
Kikibatsiogu	59	Parero	60
Kogu	78	Pauroko	116
Kogwiriru	74	Pililalo	100
Komai	20	Sulekunu	189
Lakoei	107	Tabago	119
Lamuai	80	Taguruai	83
Luilau	88	Tandareki	189
Lukauko	190	Tubaro	86
Maika	105	Turitai	64
Mogoroi	148	Ububakogu	61
Moisuru	73		

Source: Village Census 1972.

Appendix III

Trade stores by village, 1972

A. Bana Local Government Council Area

	Village	No. of trade stores		Village	No. of trade stores
Banoni Census Division			Nagovisi Census Division	Angaua	2
				Bakoram	1
				Bakupa	3
				Barario	1
North	Beteriopaia	1		Biroi	1
	Jaba	2		Kunai	1
	Koromakino	1		Labonami	1
South	Jarara	1		Lambalam	2
	Kongare	1		Lomari	1
	Mabis	4		Loro	5
	Mariga	1		Moino	1
				Mokokolio	1
Baitsi Census Division	Mewa	1		Mosino	1
				Okaru	1
				Panam	2
				Takimari	3
				Taruba	3
				Viviaro	1
				Waitabuna	2
				Waruwaru	2

B. Siwai Local Government Council Area

Village	No.	Village	No.
Aitara	1	Moronei	1
Harinai	1	Mosiraka	1
Hino	1	Mokolino (2)	1
Hire	5	Munu	2
Horino	3	Muwoko	1
Hukoha	1	Nowei	1
Iegu	1	Nukui	1
Kakatakoli	4	Panake	1
Kapana	1	Purikoli	1
Kaparo	2	Rabaru	3
Kinirui	1	Rusei	2
Kokui	2	Silaruho	1
Konga	4	Siroi	2
Kumuki	1	Sulukino	1
Kunu	1	Toitoi	1
Laku	2	Tokunoitu	2
Lakuruai	2	Tonui	7
Materas	1	Turugum	1
Matukoli	3	Ununai	1
Miheru	1	Usokoli	1

C. Buin Local Government Council Area

Census Division	Village	Number
Makis Census Division	Aku	3
	Ibiro	1
	Kogisagano	3
	Kokopo	2
	Lerot	3
	Laguai	2
	Lukararo	3
	Morula	2
	Nabuku	3
	Nakorei	2
	Piruruino	1
	Taropa	2
	Tsimbo	2
	Toaragai	1
	Tubaru	1
	Tubobisou	1
	Tugiogu	3
	Tugiu	3
	Turigau	3
Lugakei Census Division	Buruburunna	1
	Ipilai	2
	Iula	2
	Koniguru	1
	Kugugai	4
	Kukumaru	1
	Laitaro	3
	Matsiogu	1
	Moriu	2
	Okomo	1
	Pamaiuta	2
	Parerono	1
	Pariro	4
	Siul	1
Paubakei Census Division	Buin town	6
	Kanauro	8
	Kangu	1
	Kihili Mission	1
	Kikimogu	2
	Kugiogu	1
	Luaguo	1
	Malabita	5
	Mamaromino	1
	Mituai	1
	Moro	1
	Nakaro	1
	Okoiragu	1
	Turiboiru	1
	Uguimo	2
Kono Census Division	Bogisago	1
	Botulai	1
	Iamaru	1
	Kaitu	2
	Kogu	1
	Komai	2
	Lakoei	1
	Lamuai	1
	Luilau	2
	Lukauko	2
	Maika	1
	Mogoroi	2
	Moisuru	1
	Oria	1
	Orimai	2
	Pagui	4
	Pauroko	1
	Pililalo	1
	Sulekunu	3
	Tabago	5
	Tandareki	2
	Tokaino Mission	1

Source: Records of Bana, Siwai and Buin Local Government Councils.

Appendix IV

Vehicle equalisation schema

	Vehicle equivalent units
Car	1
Station wagon	1
Utility	1
Panel van	1
Trucks	1.5
Other truck types	2
Omnibus	2
Motor cycle	0.5
Tractor	1.5
Trailer	0.5
Other trailer types	2

Appendix V

The calculation of benefit-cost ratios and internal rates of return

Table AV.1

Projections of cocoa production, Siwai and Bana areas to 1981-82

Year	Siwai			Bana		
	New plantings	Mature trees	Production (tons) (at 2 lb dry bean/tree)	New plantings	Mature trees	Production (tons)
1965	-	319,699	285	-	307,241	274
1966	35,286			35,811		
1967	19,119			25,456		
1968	28,610			79,301		
1969	32,935			119,944		
1970		354,985	317		343,052	306
1971		374,104	334		368,508	329
1972		402,714	360		447,809	400
1973		435,649	389		567,753	507
1974			410			554
1975			434			621
1976			459			689
1977			483			756
1978			507			824
1979			532			891
1980			556			958
1981			580			1,026
1982			604			1,093

Table AV.2

Calculation of economic benefits for second trans-island road

Year	Siwai				Bana			
	Production (tons)			Freight	Production (tons)			Freight
	Cocoa	Copra	Total	savings ($)	Cocoa	Copra	Total	savings ($)
1971-72	360	35	395	-	400	30	430	-
1972-73	389	42	431	3,879	507	36	543	9,285
1973-74	410	50	460	4,140	554	42	596	10,192
1974-75	434	59	493	4,437	621	51	672	11,491
1975-76	459	70	529	4,761	689	60	749	12,808
1976-77	483	70	553	4,977	756	60	816	13,954
1977-78	507	70	577	5,193	824	60	884	15,116
1978-79	532	70	602	5,418	891	60	950	16,245
1979-80	556	70	626	5,634	958	60	1,018	17,408
1980-81	580	70	650	5,850	1,026	60	1,086	18,571
1981-82	604	70	674	6,066	1,093	60	1,153	19,716

Table AV.3

Calculation of total economic benefits, net benefits and present worth of net benefits for present trans-island road

Year	Savings ($)				Total economic benefits ($)	Maintenance costs ($)	Net benefits ($)	Present worth at 20%
	Outwards freight		Inwards freight					
	Siwai	Bana	Siwai	Bana				
1972-73	3,879	9,285	10,456	24,310	47,930	14,000	33,930	28,275
1973-74	4,140	10,192	12,314	26,134	52,780	14,000	38,780	26,931
1974-75	4,437	11,491	13,238	28,094	57,260	14,000	43,260	25,035
1975-76	4,761	12,808	14,230	30,200	61,999	14,000	47,999	23,148
1976-77	4,977	13,954	15,298	32,466	66,695	14,000	52,695	21,177
1977-78	5,193	15,116	15,910	33,764	69,983	14,000	55,983	18,749
1978-79	5,418	16,245	16,546	35,116	73,325	14,000	59,325	16,557
1979-80	5,634	17,408	17,208	36,520	76,770	14,000	62,770	14,598
1980-81	5,850	18,571	17,898	37,980	80,299	14,000	66,299	12,849
1981-82	6,066	19,716	18,612	39,500	83,894	14,000	69,894	11,288
								198,607

Table AV.4

Calculation of economic benefits for new trans-island road

Year	Bana				Freight savings (x S5.74)	Siwai				Freight savings (x $4.58)	Buin				Freight savings (x $3.71)	Total economic benefits ($)
	Production (tons)		Inwards freight (tons)	Total (tons)		Production (tons)		Inwards freight (tons)	Total (tons)		Production (tons)		Inwards freight (tons)	Total (tons)		
	Cocoa	Copra				Cocoa	Copra				Cocoa	Copra				
1977-78	824	60	1,700	2,584	14,832	507	70	1,700	2,277	10,429	957	106	1,700	2,763	10,251	35,512
1978-79	891	60	1,768	2,719	15,607	532	70	1,768	2,370	10,855	1,019	106	1,768	2,893	10,733	37,195
1979-80	958	60	1,838	2,856	16,393	556	70	1,838	2,464	11,285	1,190	106	1,838	3,134	11,627	39,305
1980-81	1,026	60	1,912	2,998	17,209	580	70	1,912	2,562	11,734	1,291	106	1,912	3,309	12,276	41,219
1981-82	1,093	60	1,989	3,142	18,035	604	70	1,989	2,663	12,197	1,393	106	1,989	3,488	12,940	43,172
1982-83	1,160	60	2,068	3,288	18,873	628	70	2,068	2,766	12,668	1,494	106	2,068	3,668	13,608	45,149
1983-84	1,228	60	2,151	3,439	19,740	652	70	2,151	2,873	13,158	1,596	106	2,151	3,853	14,295	47,193
1984-85	1,295	60	2,237	3,592	20,618	676	70	2,237	2,983	13,662	1,697	106	2,237	4,040	14,988	49,268
1985-86	1,362	60	2,327	3,749	21,519	701	70	2,327	3,098	14,189	1,798	106	2,327	4,231	15,697	51,405
1986-87	1,430	60	2,420	3,910	22,443	725	70	2,420	3,215	14,725	1,900	106	2,420	4,426	16,420	53,588

Table AV.5

Calculation of social travel benefits for new trans-island road

Year	Bana		Siwai		Buin		Total social travel benefits	Less 10%
	Population	Social travel benefits ($)	Population	Social travel benefits ($)	Half population	Social travel benefits ($)	($)	($)
1977-78	10,154	8,296	10,413	14,276	7,924	16,640	39,212	35,291
1978-79	10,560	12,941	10,830	22,272	8,241	25,959	61,172	55,055
1979-80	10,983	17,946	11,263	30,883	8,571	35,998	84,827	76,344
1980-81	11,422	23,329	11,713	40,146	8,913	46,793	110,268	99,241
1981-82	11,879	29,115	12,812	52,696	9,270	58,401	140,212	126,191
1982-83	12,354	35,326	12,669	60,792	9,641	70,861	166,979	150,281
1983-84	12,848	41,987	13,176	72,257	10,026	84,218	198,462	178,616
1984-85	13,362	49,125	13,703	84,541	10,427	98,535	232,201	208,981
1985-86	13,896	56,765	14,251	97,691	10,845	113,873	268,329	241,496
1986-87	14,452	59,036	14,821	101,598	11,278	118,419	279,053	251,148

Table AV.6

Present worth of maximum construction costs

Year		Present worth			
		at 5%	at 10%	at 15%	at 20%
	($)	($)	($)	($)	($)
1974-75	380,625	362,500	346,023	330,978	317,188
1975-76	761,250	690,476	629,132	575,614	528,646
1976-77	380,625	328,798	285,969	250,267	220,269
Total	1,522,500	1,381,774	1,261,124	1,156,859	1,066,103

Present worth of minimum construction costs

Year		Present worth			
		at 5%	at 10%	at 15%	at 20%
	($)	($)	($)	($)	($)
1974-75	217,375	207,024	197,614	189,022	181,146
1975-76	434,750	394,331	359,298	328,733	301,910
1976-77	217,375	187,777	163,317	142,928	125,796
Total	869,500	789,132	720,229	660,683	608,852

Table AV.7

Calculation of present worth of net benefits for new trans-island road

Year	Economic benefits	Social travel benefits	Total benefits	Maintenance costs	Net benefits	Present worth			
						at 5%	at 10%	at 15%	at 20%
	($)	($)	($)	($)	($)	($)	($)	($)	($)
1977-78	35,512	35,291	70,803	16,125	54,678	44,984	37,346	31,262	26,369
1978-79	37,195	55,055	92,250	16,125	76,125	59,646	47,268	37,848	30,593
1979-80	39,305	76,344	115,649	16,125	99,524	74,266	56,179	43,027	33,330
1980-81	41,219	99,241	140,460	16,125	124,335	88,363	63,804	46,742	34,700
1981-82	43,172	126,191	169,363	16,125	153,238	103,718	71,487	50,094	35,638
1982-83	45,149	150,281	195,430	16,125	179,305	115,582	76,043	50,970	34,751
1983-84	47,193	178,616	225,809	16,125	209,684	128,728	80,842	51,831	33,865
1984-85	49,268	208,981	258,249	16,125	242,124	141,565	84,863	52,043	32,587
1985-86	51,405	241,496	292,901	16,125	276,776	154,119	88,189	51,731	31,042
1986-87	53,588	251,148	304,736	16,125	288,611	153,057	83,600	46,907	26,975
					1,704,400	1,064,028	689,621	462,455	319,850

Table AV.8

Projections of cocoa production, Buin area, to 1986-87

Year	New plantings	Total (mature and immature)	Mature	Production (tons) estimated (at 2 lb dry bean/tree)
1965		261,743		
1965-66	41,773			
1966-67	50,578			
1967-68	154,839			
1968-69	186,118			
1969-70	[123,889 est.]			
1970-71	[123,889 est.]	942,828	303,516	271
1971-72	98,403		354,094	316
1972-73	30,231		508,933	454
1973-74	70,065		695,051	621
1974-75			818,940	731
1975-76			942,829	842
1976-77			1,041,232	930
1977-78			1,071,463	957
1978-79			1,141,528	1,019
1979-80				1,190
1980-81				1,291
1981-82				1,393
1982-83				1,494
1983-84				1,596
1984-85				1,697
1985-86				1,798
1986-87				1,900
				2,001

Table AV.9

Calculation of economic benefits for Buin-Toimonapu road

Year	Buin Production (tons) Cocoa	copra	Inwards freight (tons)	Total (tons)	Freight savings (x $9.40) ($)
1978-79	1,019	106	1,768	2,893	27,194
1979-80	1,190	106	1,838	3,134	29,460
1980-81	1,291	106	1,912	3,309	31,105
1981-82	1,393	106	1,989	3,488	32,787
1982-83	1,494	106	2,068	3,668	34,479
1983-84	1,596	106	2,151	3,853	36,218
1984-85	1,697	106	2,237	4,040	37,976
1985-86	1,798	106	2,327	4,231	39,771
1986-87	1,900	106	2,420	4,426	41,604
1987-88	2,001	106	2,517	4,624	43,466

Table AV.10

Calculation of social travel benefits for Buin-Toimonapu road

Year	Buin population	Social travel savings ($)	Less 10% ($)
1978-79	16,482	23,075	20,767
1979-80	17,142	35,998	32,938
1980-81	17,826	49,913	44,922
1981-82	18,540	64,890	58,392
1982-83	19,282	80,984	72,886
1983-84	20,052	98,255	88,429
1984-85	20,854	116,782	105,104
1985-86	21,690	136,647	122,982
1986-87	22,556	157,892	142,103
1987-88	23,458	164,206	147,785

Table AV.11

Present worth of minimum cost, Buin-Toimonapu road

Year	Construction costs ($)	Present worth at 5% ($)
1974-75	425,750	386,168
1975-76	851,500	735,558
1976-77	425,750	350,266
	1,703,000	1,471,992

Table AV.12

Calculation of present worth of benefits and costs, Buin-Toimonapu road

Year	Economic benefits ($)	Social travel benefits ($)	Total benefits ($)	Maintenance costs ($)	Net benefits ($)	Present worth at 5% ($)
1978-79	27,194	20,767	47,961	24,725	23,236	18,206
1979-80	29,460	32,938	62,398	24,725	37,673	28,112
1980-81	31,105	44,922	76,027	24,725	51,302	36,459
1981-82	32,787	58,392	91,179	24,725	66,454	44,979
1982-83	34,479	72,886	107,365	24,725	82,640	53,270
1983-84	36,218	88,429	124,647	24,725	99,922	61,343
1984-85	37,976	105,104	143,080	24,725	118,355	69,200
1985-86	39,771	122,982	162,753	24,725	138,028	76,859
1986-87	41,604	142,103	183,707	24,725	158,982	84,312
1987-88	43,466	147,785	191,251	24,725	166,526	84,107
					836,598	556,847

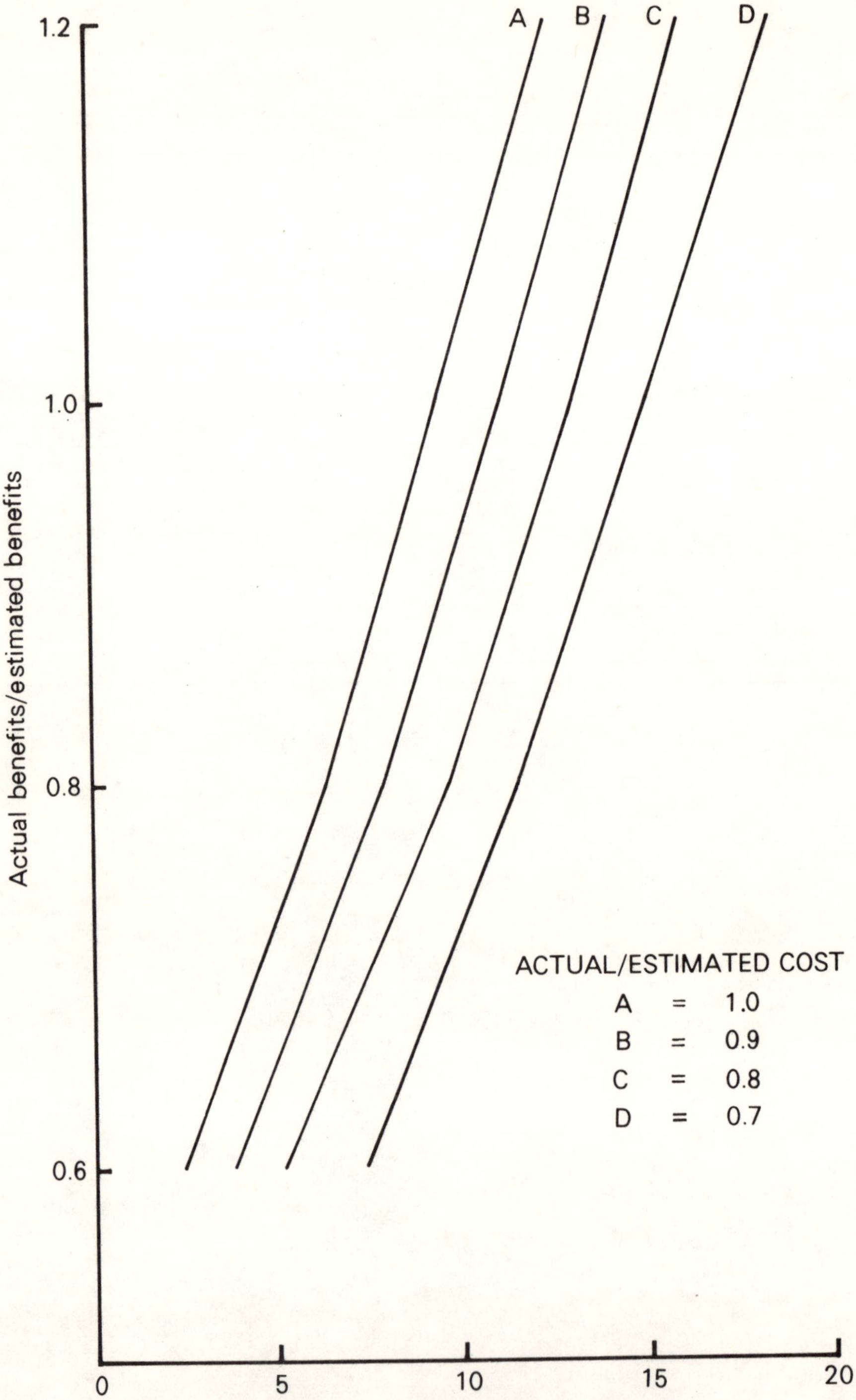

Figure A5.1 Sensitivity analysis, internal rate of return, new trans-island road

Sensitivity analysis of new trans-island road

Minimum cost case

Present worth at discount rate of	Actual/estimated cost ($) (A) =1.0	(B) =0.9	(C) =0.8	(D) =0.7
0%	869,500	782,550	695,600	608,650
5%	789,132	710,219	631,306	552,392
10%	720,229	648,206	576,183	504,160
15%	660,683	594,615	528,546	462,478
20%	608,852	547,967	487,082	426,196

Case 1. Actual/estimated benefits = 1.2

Present worth of benefits ($)

at 0% = 2,045,280
5% = 1,276,834
10% = 827,545
15% = 554,946
20% = 383,820

(a) If actual/estimated cost = A=1.0

$$\text{BCR at } 10\% = \frac{827,545}{720,229} = 1.15$$

$$15\% = \frac{554,946}{660,683} = 0.84$$

IRR = 12.42%

(b) If actual/estimated cost = B=0.9

$$\text{BCR at } 10\% = \frac{827,545}{648,206} = 1.28$$

$$15\% = \frac{554,946}{594,615} = 0.93$$

IRR = 14.00%

(c) If actual/estimated cost = C=0.8

$$\text{BCR at } 15\% = \frac{554,946}{528,546} = 1.05$$

$$20\% = \frac{383,820}{487,082} = 0.79$$

IRR = 15.96%

(d) If actual/estimated cost = D=0.7

$$\text{BCR at } 15\% = \frac{554,946}{462,478} = 1.20$$

$$20\% = \frac{383,820}{426,196} = 0.90$$

IRR = 18.33%

Case 2. Actual/estimated benefits = 1.0

Present worth of benefits ($)

at 0% = 1,704,400
5% = 1,064,028
10% = 689,621
15% = 462,455
20% = 319,850

(a) If actual/estimated cost = A=1.0

$$\text{BCR at } 5\% = \frac{1,064,028}{789,132} = 1.35$$

$$10\% = \frac{689,621}{720,229} = 0.96$$

IRR = 9.38%

(b) If actual/estimated cost = B=0.9

$$\text{BCR at } 10\% = \frac{689,621}{648,206} = 1.06$$

$$15\% = \frac{462,455}{594,615} = 0.78$$

IRR = 11.07%

(c) If actual/estimated cost = C=0.8

$$\text{BCR at } 10\% = \frac{689,621}{576,183} = 1.20$$

$$15\% = \frac{462,455}{528,546} = 0.87$$

IRR = 13.03%

(d) If actual/estimated cost = D=0.7

$$\text{BCR at } 10\% = \frac{689,621}{504,160} = 1.37$$

$$15\% = \frac{462,455}{462,478} = 0.99995$$

IRR = 14.999%

Case 3. Actual/estimated benefits = 0.8

Present worth of benefits

	($)
at 0% =	1,363,520
5% =	851,222
10% =	551,697
15% =	369,964

(a) If actual/estimated cost = A=1.0

$$\text{BCR at } 5\% = \frac{851,222}{789,132} = 1.08$$

$$10\% = \frac{551,697}{720,229} = 0.77$$

IRR = 6.29%

(b) If actual/estimated cost = B=0.9

$$\text{BCR at } 5\% = \frac{851,222}{710,219} = 1.20$$

$$10\% = \frac{551,697}{648,206} = 0.85$$

IRR = 7.86%

(c) If actual/estimated cost = C=0.8

$$\text{BCR at } 5\% = \frac{851,222}{631,306} = 1.35$$

$$10\% = \frac{551,697}{576,183} = 0.96$$

IRR = 9.62%

(d) If actual/estimated cost = D=0.7

$$\text{BCR at } 10\% = \frac{551,697}{504,160} = 1.09$$

$$15\% = \frac{369,964}{462,478} = 0.80$$

IRR = 11.55%

Case 4. Actual/estimated benefits = 0.6

Present worth of benefits

	($)
at 0% =	1,022,640
5% =	638,417
10% =	413,773
15% =	277,473
20% =	191,910

(a) If actual/estimated cost = A=1.0

$$\text{BCR at } 0\% = \frac{1,022,640}{869,500} = 1.18$$

$$5\% = \frac{638,417}{789,132} = 0.81$$

IRR = 2.43%

(b) If actual/estimated cost = B=0.9

$$\text{BCR at } 0\% = \frac{1,022,640}{782,550} = 1.31$$

$$5\% = \frac{638,417}{710,219} = 0.90$$

IRR = 3.78%

(c) If actual/estimated cost = C=0.8

$$\text{BCR at } 5\% = \frac{638,417}{631,306} = 1.01$$

$$10\% = \frac{413,773}{576,183} = 0.72$$

IRR = 5.17%

(d) If actual/estimated cost = D=0.7

$$\text{BCR at } 5\% = \frac{638,417}{552,392} = 1.16$$

$$10\% = \frac{413,773}{504,160} = 0.82$$

IRR = 7.35%

These results are shown graphically in Fig.AV.1.

Calculation of benefit-cost ratios for feeder roads or circuits

Schedule 1 Nagovisi-Baitsi

	Biroi	Lomari	Angaua	Aumari	Mingeta	Nukui	Osiangi	Waruwaru	Bererekei	Momogonari	Waitabuna	Loro	Lopere	Sovele	Total
No. of cocoa trees, 1970	-	20,112	19,960	6,587	2,146	21,802	38,777	19,413(½)	5,895(½)	8,433(½)	18,302	52,631	16,528	-	230,586
Population 1972	-	135	142	108	48	180	191	126	66	51	124	305	136	-	1,612

Length of road (miles) 14½ (less 2 shared to next circuit, so say 13½)

Economic benefits start at $\$2 \times \frac{230,586}{373} \times 2 \times 13.5 \times \frac{9.5}{100} + \frac{3}{10} \times \frac{230,586}{373} \times 44.80 = 3,171 + 8,309 = \$11,480$

Social travel benefits start at $\$ \frac{1,612}{25} \times 50 \times 2 \times 13.5 \times \frac{9.5}{100} = \$8,270$

Costs:

Construction 13½ miles at $5,000/mile = $67,500

Maintenance at $300/mile = $4,050

Year	Economic benefits (growth at 6% p.a.) ($)	Social travel benefits (growth at 4% p.a.) ($)	Total benefits ($)	Less maintenance costs ($)	Net benefits ($)	Present worth at 5% ($)	10% ($)	15% ($)	20% ($)
1970	11,480								
1971	-								
1972		8,270							
1973									
1974									
1975									
1976	16,285	9,675	25,960	4,050	21,910	20,867	19,918	19,052	18,258
1977	17,262	10,062	27,324	4,050	23,274	21,110	19,235	17,598	16,163
1978	18,297	10,464	28,761	4,050	24,711	21,346	18,566	16,248	14,300
1979	19,395	10,883	30,278	4,050	26,228	21,578	17,914	14,996	12,649
1980	20,559	11,318	31,877	4,050	27,827	21,803	17,278	13,835	11,183
1981	21,792	11,771	33,563	4,050	29,513	22,023	16,659	12,759	9,884
1982	23,100	12,242	35,342	4,050	31,292	22,239	16,058	11,764	8,733
1983	24,486	12,731	37,217	4,050	33,167	22,449	15,473	10,842	7,714
1984	25,955	13,241	39,196	4,050	35,146	22,655	14,481	9,991	6,812
1985	27,512	13,770	41,282	4,050	37,232	22,857	14,355	9,203	6,013
						218,927	169,837	136,288	111,709

Benefit-cost ratio at 5% = 32 10% = 2.5 15% = 2.0 20% = 1.7

Internal rate of return = 35%

Schedule 2 Nagovisi-Baitsi

	Momogonari (½)	Berereke (½)	Waruwaru (½)	Sianeki	Barario	Labonami	Panam	Lambalam	Sikoreva	Koro	Puranavia	Leira	Borioko	Total
No. of cocoa trees, 1970	8,433	5,895	19,413	10,910	7,426	8,753	28,317	41,884	10,386	20,233	14,764	22,162	7,191	205,767
Population 1972	51	66	126	141	144	153	177	287	115	195	118	141	108	1,822

Length of road (miles) 18 (less 2 shared, say 16)

Economic benefits start at $\$2 \times \frac{205,767}{373} \times 2 \times 18 \times \frac{9.5}{100} + \frac{3}{10} \times \frac{205,767}{373} \times 44.80 = \$3,773 + 7,414 = \$11,187$

Social travel benefits start at $\$ \frac{1,822}{25} \times 50 \times 2 \times 18 \times \frac{9.5}{100} = \$12,462$

Costs:

Construction 16 miles at \$5,000/mile = \$80,000

Maintenance at \$300/mile = \$4,800 p.a.

Year	Economic benefits (growth at 6% p.a.) ($)	Social travel benefits (growth at 4% p.a.) ($)	Total benefits ($)	Less maintenance costs ($)	Net benefits ($)	Present worth at 5% ($)	10% ($)	15% ($)	20% ($)
1970	11,187								
1971									
1972		12,462							
1973									
1974									
1975									
1976	15,869	14,579	30,448	4,800	25,648	24,427	23,316	22,303	21,373
1977	16,821	15,162	31,983	4,800	27,183	24,656	22,465	20,554	18,877
1978	17,830	15,768	33,598	4,800	28,798	24,877	21,636	18,935	16,666
1979	18,900	16,399	35,299	4,800	30,499	25,092	20,831	17,438	14,708
1980	20,034	17,055	37,089	4,800	32,289	25,299	20,049	16,053	12,976
1981	21,236	17,737	38,973	4,800	34,173	25,500	19,290	14,774	11,444
1982	22,510	18,447	40,957	4,800	36,157	25,696	18,554	13,593	10,091
1983	23,861	19,185	43,046	4,800	38,246	25,886	17,842	12,503	8,895
1984	25,293	19,952	45,245	4,800	40,445	26,071	17,153	11,497	7,839
1985	26,810	20,750	47,560	4,800	42,760	26,251	16,486	10,570	6,906
						253,755	197,622	158,220	129,775

Benefit-cost ratio at 5% = 3.2 10% = 2.5 15% = 2.0 20% = 1.6

Internal rate of return = 34%

Schedule 3 Nagovisi-Baitsi

Year	Boku	Pikei	Biros	Bakoram 1	Bakoram 2	Pomaletti	Biroi	Sovele	Total
No. of cocoa trees, 1970	10,422	25,404	34,418	40,054	15,946	24,106	16,059		166,409
Population 1972	147	257	177	248	197	202	105		1,333
Length of road (miles)	8½								

Economic benefits start at $\$2 \times \frac{166,409}{373} \times 2 \times 8.5 \times \frac{9.5}{100} + \frac{3}{10} \times \frac{166,409}{373} \times 44.80 = \$1,441 + 5,996 = \$7,437$

Social travel benefits start at $\$ \frac{1,333}{25} \times 50 \times 248.5 \times \frac{9.5}{100} = \$4,306$

Costs:

Construction 8½ miles at \$5,000/mile = \$42,500

Maintenance at \$300/mile = \$2,550

Year	Economic benefits (growth at 6% p.a.) ($)	Social travel benefits (growth at 4% p.a.) ($)	Total benefits ($)	Less maintenance costs ($)	Net benefits ($)	Present worth at 5% ($)	Present worth at 10% ($)	Present worth at 15% ($)	Present worth at 20% ($)
1970	7,437								
1971									
1972		4,306							
1973									
1974									
1975									
1976	10,550	5,037	15,587	2,550	13,037	12,416	11,852	11,337	10,864
1977	11,183	5,239	16,422	2,550	13,872	12,582	11,464	10,489	9,633
1978	11,710	5,448	17,158	2,550	14,608	12,619	10,975	9,605	8,141
1979	12,565	5,666	18,321	2,550	15,681	12,901	10,710	8,966	7,562
1980	13,319	5,893	19,212	2,550	16,662	13,055	10,346	8,284	6,696
1981	14,118	6,129	20,247	2,550	17,697	13,206	9,990	7,651	5,927
1982	14,965	6,374	21,339	2,550	18,789	13,353	9,642	7,063	5,244
1983	15,863	6,629	22,492	2,550	19,942	13,498	9,303	6,519	4,638
1984	16,814	6,894	23,708	2,550	21,158	13,639	8,973	6,014	4,101
1985	17,823	7,170	24,993	2,550	22,443	13,778	8,653	5,548	3,625
						131,047	101,908	81,476	66,431

Benefit-cost ratio 5% = 3.1 10% = 2.4 15% = 1.9 20% = 1.6

Internal rate of return = 33%

Schedule 4 Nagovisi-Baitsi

	Bakupa	Moino	Okaru	Sipi	Taguri	Lotari 1	Lotari 2	Total
No. of cocoa trees, 1970	26,445	9,671	9,581	14,596	8,517	21,575	2,844	93,229
Population 1972	160	130	184	189	101	123	138	1,025

Length of road (miles) 8

Economic benefits start at $\$2 \times \frac{93,229}{373} \times 2 \times 8 \times \frac{9.5}{100} + \frac{3}{10} \times \frac{93,229}{373} \times 44.80 = \$760 + 3,359 = \$4,119$

Social travel benefits start at $\$ \frac{1,025}{25} \times 50 \times 2 \times 8 \times \frac{9.5}{100} = \$3,116$

Costs:

Construction 8 miles at $5,000/mile = $40,000

Maintenance at $300/mile = $2,400 p.a.

Year	Economic benefits (growth at 6% p.a.) ($)	Social travel benefits (growth at 4% p.a.) ($)	Total benefits ($)	Less maintenance costs ($)	Net benefits ($)	Present worth at 5% ($)	10% ($)	15% ($)	20% ($)
1970	4,119								
1971									
1972		3,116							
1973									
1974									
1975									
1976	5,843	3,645	9,488	2,400	7,088	6,750	6,444	6,163	5,907
1977	6,193	3,791	9,984	2,400	7,584	6,879	6,268	5,735	5,267
1978	6,565	3,943	10,508	2,400	8,108	7,004	6,092	5,331	4,692
1979	6,959	4,100	11,059	2,400	8,659	7,124	5,914	4,951	4,176
1980	7,377	4,264	11,641	2,400	9,241	7,241	5,738	4,594	3,714
1981	7,819	4,435	12,254	2,400	9,854	7,353	5,562	4,260	3,300
1982	8,288	4,612	12,900	2,400	10,500	7,462	5,388	3,947	2,930
1983	8,786	4,797	13,583	2,400	11,183	7,569	5,217	3,656	2,601
1984	9,313	4,989	14,302	2,400	11,902	7,672	5,048	3,383	2,307
1985	9,871	5,188	15,059	2,400	12,659	7,772	4,881	3,129	2,045
						72,826	56,552	45,149	36,939

Benefit-cost ratio at 5% = 1.8 10% = 1.4 15% = 1.1 20% = 0.9.

Internal rate of return = 18.5%

Schedule 5 Nagovisi-Baitsi-South Banoni

	Mariga	Kongare	Mabis	Jarara	Total
No. of cocoa trees, 1970	3,730	11,445	11,391	10,805	37,371
				There are 25 tons p.a. of copra from this area. Farm gate price for copra = $127/ton.	
Population 1972	82	126	132	188	528
Length of road (miles)	6.5				

Economic benefits start at $\$2 \times \frac{37,371}{373} \times 2 \times 6.5 \times \frac{18.9}{100} + \frac{3}{10} \times \frac{37,371}{373} \times 44.80 + 2 \times 25 \times 13 \times \frac{18.9}{100} + \frac{30}{100} \times 2 \times 25 \times 127 = \$492 + 1,347 + 123 + 1,905 = \$1,839 + 2,028$

Social travel benefits start at $\$\ \frac{528}{25} \times 50 \times 2 \times 6.5 \times \frac{18.9}{100} = \$2,595$

Costs:

Construction 6.5 miles at $5,000/mile = $32,500

Maintenance 6.5 miles at $300/mile = $1,950 p.a.

Year	Economic benefits Cocoa ($)	Economic benefits Copra ($)	Social travel benefits (growth at 4% p.a.) ($)	Total benefits ($)	Less maintenance costs ($)	Net benefits ($)	Present worth at 5% ($)	Present worth at 10% ($)	Present worth at 15% ($)	Present worth at 20% ($)
1970	1,839	2,028								
1971		2,028								
1972		2,028	2,595							
1973		2,028								
1974		2,028								
1975		2,028								
1976	2,609	2,028	3,036	7,673	1,950	5,723	5,450	5,203	4,977	4,769
1977	2,765	2,028	3,157	7,950	1,950	6,000	5,442	4,959	4,537	4,167
1978	2,931	2,028	3,284	8,243	1,950	6,293	5,436	4,728	4,138	3,642
1979	3,107	2,028	3,415	8,550	1,950	6,600	5,430	4,508	3,774	3,183
1980	3,293	2,028	3,551	8,872	1,950	6,922	5,424	4,298	3,441	2,782
1981	3,491	2,028	3,693	9,212	1,950	7,262	5,419	4,099	3,140	2,432
1982	3,700	2,028	3,841	9,569	1,950	7,619	5,415	3,910	2,864	2,126
1983	3,922	2,028	3,995	9,945	1,950	7,995	5,411	3,730	2,614	1,859
1984	4,158	2,028	4,155	10,341	1,950	8,391	5,409	3,559	2,385	1,626
1985	4,407	2,028	4,321	10,756	1,950	8,806	5,406	3,395	2,177	1,422
							54,242	42,389	34,047	28,008

Benefit-cost ratio at 5% = 1.7 10% = 1.3 15% = 1.05 20% = 0.86

Internal rate of return = 16%

Schedule 6 Nagovisi-Baitsi

	(a) Taruba	Agabei	Mosino	Masawakori		(b) Moratona	Kuinai	(c) Mosigetta	1½m west	Total
No. of cocoa trees, 1970	22,778	9,569	-	10,396	(42,743)	-	10,847	-	20,000	73,590
Population 1972	141	109	-	92			91			433

Length of road (miles) $3 + 1\frac{1}{2} + 1\frac{1}{2} = 6$

Economic benefits start at $\$2 \times \frac{73,590}{373} \times 2 \times 6 \times \frac{9.5}{100} + 3 \times \frac{73,590}{373} \times 44.80 = \$450 + 2,652 = \$3,102$

Social travel benefits start at $\$ \frac{433}{25} \times 50 \times 2 \times 6 \times \frac{9.5}{100} = \$987$

Costs:

Construction 6 miles at \$5,000/mile = \$30,000

Maintenance at \$300/mile = \$1,800

Year	Economic benefits (growth at 6% p.a.) ($)	Social travel benefits (growth at 4% p.a.) ($)	Total benefits ($)	Less maintenance costs ($)	Net benefits ($)	Present worth at 5% ($)	Present worth at 10% ($)	Present worth at 15% ($)	Present worth at 20% ($)
1970	3,102								
1971									
1972		987							
1973									
1974									
1975									
1976	4,400	1,155	4,555	1,800	3,755	3,576	3,414		
1977	4,664	1,201	5,865	1,800	4,065	3,687	3,360		
1978	4,944	1,249	6,193	1,800	4,393	3,795	3,301		
1979	5,241	1,299	6,540	1,800	4,740	3,900	3,237		
1980	5,555	1,351	6,906	1,800	5,106	4,001	3,170		
1981	5,889	1,405	7,294	1,800	5,494	4,100	3,102		
1982	6,242	1,461	7,703	1,800	5,903	4,195	3,029		
1983	6,616	1,519	8,135	1,800	6,335	4,288	2,955		
1984	7,013	1,580	8,593	1,800	6,793	4,379	2,881		
1985	7,434	1,643	9,077	1,800	7,277	4,467	2,806		
						40,388	28,226		

Benefit-cost ratio at 5% = 1.35 10% = 0.94

Internal rate of return = 9%

Schedule 7 Siwai

	Konga	Kutin	Kokui	Kunu	Misiminoi	Rusei	Moibinei	Lakemba	Suluhino	Harinai	Harikagu	Ununai	Hire	Koropo	Toitoi	Monoitu R.C.M.	Amio	Siroi	Osokori	Total
No. of cocoa trees, 1970	-	3,935	9,959	13,370	18,580	1,048	4,237	2,400	3,040	8,478	2,450	7,306	9,632	3,360	8,350	-	3,951	2,956	4,350	107,402
Population 1972	-	97	157	164	198	308	92	116	137	165	80	162	201	73	124	-	186	192	81	2,533
Length of road (miles)			12																	

Economic benefits start at $\$2 \times \frac{107,402}{373} \times 2 \times \frac{18.9 \times 12}{100} + \frac{3}{10} \times \frac{107,402}{373} \times 44.80 = \$2,612 + 3,870 = \$6,482$

Social travel benefits start at $\$ \frac{2,533}{25} \times 50 \times 2 \times \frac{18.9 \times 12}{100} = \$22,979$

Costs:

Construction 12 miles at $5,000/mile = $60,000

Maintenance at $300/mile = $3,600

Year	Economic benefits (growth at 6% p.a.) ($)	Social travel benefits (growth at 4% p.a.) ($)	Total benefits ($)	Less maintenance costs ($)	Net benefits ($)	Present worth at 5% ($)	10% ($)	15% ($)	20% ($)
1970	6,482								
1971									
1972		22,979							
1973									
1974									
1975									
1976	9,747	27,957	37,704	3,600	34,104	32,480	31,004	29,656	28,420
1977	10,331	29,076	39,407	3,600	35,807	32,478	29,593	27,075	24,866
1978	10,951	32,239	41,190	3,600	37,590	32,472	28,242	24,716	21,753
1979	11,608	31,448	43,056	3,600	39,456	32,461	26,949	22,559	19,028
1980	12,305	32,706	45,011	3,600	41,411	32,460	25,713	20,589	16,642
1981	13,043	34,015	47,058	3,600	43,458	32,429	24,531	18,788	14,554
1982	13,826	35,375	49,201	3,600	45,601	32,408	23,401	17,143	12,726
1983	14,655	36,790	51,445	3,600	47,845	32,383	22,320	15,641	11,127
1984	15,534	38,262	53,796	3,600	50,196	32,357	21,288	14,269	9,728
1985	16,467	39,792	56,259	3,600	52,659	32,328	20,302	13,017	8,505
						324,256	253,343	203,453	167,349

Benefit-cost ratio at 5% = 5.4 10% = 4.2 15% = 3.4 20% = 2.8

Internal rate of return = 61%

Schedule 8 Siwai

	Konga	Iegu	Materas	Kuhino	Hino	Taropa	Laku	Moronei	Maintoita	Moisua	Hanong (Materas)	Naronai	Morokaimoro	Mokolino 1 and 2	Rokuse	Turugum + 4m inland	Total
No. of cocoa trees, 1970	-	4,937	9,926	2,000	4,450	15,299	4,760	10,003	3,800	7,310	3,300	1,970	3,361	7,301	300	9,323	88,040
Population 1972	-	86	100	112	120	109	120	182	83	117	93	73	149	212	54	192	1,802

Length of road (miles): 14 miles existing plus 4 miles new (for which benefits not included)

Economic benefits start at: $\$2 \times \frac{88,040}{373} \times 2 \times \frac{18.9 \times 14}{100} + \frac{3}{10} \times \frac{88,040}{373} \times 44.80 = \$2,498 + 3,172 = \$5,760$

Social travel benefits start at: $\$ \frac{1,802}{25} \times 50 \times 2 \times \frac{18.9 \times 14}{100} = \$19,072$

Costs:

Construction: 18 miles at $5,000/mile = $90,000

Maintenance: 18 miles at $300/mile = $5,400 p.a.

Year	Economic benefits (growth at 6% p.a.) ($)	Social travel benefits (growth at 4% p.a.) ($)	Total benefits ($)	Less maintenance costs ($)	Net benefits ($)	Present worth at 5% ($)	10% ($)	15% ($)	20% ($)
1970	5,760								
1971									
1972		19,072							
1973									
1974									
1975									
1976	8,661	23,204	31,865	5,400	26,465	25,205	24,059	23,013	22,054
1977	9,181	24,132	33,313	5,400	27,913	25,318	23,069	21,106	19,384
1978	9,731	25,097	34,828	5,400	29,428	25,421	22,110	19,349	17,030
1979	10,315	26,101	36,416	5,400	31,016	25,517	21,184	17,734	14,958
1980	10,934	27,145	38,079	5,400	32,679	25,605	20,291	16,247	13,133
1981	11,590	28,231	39,821	5,400	34,421	25,685	19,430	14,881	11,528
1982	12,286	29,360	41,646	5,400	36,246	25,759	18,600	13,626	10,116
1983	13,023	30,535	43,558	5,400	38,158	25,827	17,801	12,474	8,874
1984	13,804	31,756	45,560	5,400	40,160	25,887	17,032	11,416	7,783
1985	14,632	33,027	47,659	5,400	42,259	25,943	16,293	10,446	6,825
						256,167	178,685	160,292	131,685

Benefit-cost ratio at 5% = 2.9 10% = 2.0 15% = 1.8 20% = 1.46

Internal rate of return = 31%

Schedule 9 Siwai

	Konga	Kinirui	Munu	Sikurai	Matukoli	Rabaru	Kupingku	Silaruho	Purikoli	Kakatakoli + 2 miles inland	Total
No. of cocoa trees, 1970	-	4,320	4,600	3,652	13,726	7,800	2,622	2,400	6,400	7,205	52,725
Population 1972	-	206	208	54	218	202	83	77	103	147	1,298

Length of road (miles) 10.5 miles of which 4 new (includes 2 miles inland for which benefits are not counted).

Economic benefits start at $\$2 \times \frac{52,725}{373} \times 2 \times \frac{18.9 \times 6.5 + 9.5 \times 1.5}{100} + \frac{3}{10} \times \frac{52,725}{373} \times 44.80 = \frac{122.85 + 11.0}{100} = 1.3385 = \$757 + 1,900 = \$2,657$

Social travel benefits start at $\$\ \frac{1,298}{25} \times 50 \times 2(1.3385) = \$6,949$

Costs:

Construction 10.5 miles at $5,000/mile = $52,500

Maintenance at $300/mile = $3,150

Year	Economic benefits (growth at 6% p.a.) ($)	Social travel benefits (growth at 4% p.a.) ($)	Total benefits ($)	Less maintenance costs ($)	Net benefits ($)	Present worth at 5% ($)	10% ($)	15% ($)	20% ($)
1970	2,657								
1971									
1972		6,949							
1973									
1974									
1975									
1976	3,995	8,455	12,450	3,150	9,300	8,857	8,455	8,087	7,750
1977	4,235	8,793	13,028	3,150	9,878	8,960	8,164	7,469	6,860
1978	4,489	9,144	13,633	3,150	10,483	9,056	7,876	6,893	6,067
1979	4,758	9,510	14,268	3,150	11,118	9,147	7,594	6,357	5,362
1980	5,044	9,891	14,935	3,150	11,785	9,234	7,318	5,859	4,736
1981	5,346	10,286	15,632	3,150	12,482	9,314	5,258	5,396	4,180
1982	5,667	10,698	16,365	3,150	13,215	9,392	4,820	4,968	3,688
1983	6,007	11,126	17,133	3,150	13,983	9,464	4,415	4,571	3,252
1984	6,368	11,571	17,939	3,150	14,789	9,533	4,043	4,204	2,710
1985	6,750	12,033	18,783	3,150	15,633	9,597	3,700	3,864	2,525
						92,554	61,643	57,668	47,130

Benefit-cost ratio at 5% = 1.76 10% = 1.17 15% = 1.10 20% = 0.90

Internal rate of return = 17.0%

Schedule 10 Siwai

	Horino	Kimaku	Tokunoitu	Haisi	Hiru Hiru	Total
No. of cocoa trees, 1970	-	4,860	7,414	8,840	8,900	30,014
Population 1972	-	105	139	134	146	524
Length of road (miles)	6					

Economic benefits start at $\$2 \times \frac{30,014}{373} \times 2 \times \frac{18.9 \times 6}{100} + \frac{3}{10} \times \frac{30,014}{373} \times 44.80 = \$365 + 1,081 = \$1,446$

Social travel benefits start at $\$ \frac{524}{25} \times 50 \times 2 \times \frac{18.9 \times 6}{100} = \$2,377$

Costs:

Construction 6 miles at $5,000/mile = $30,000

Maintenance 6 miles at $300/mile = $1,800

Year	Economic benefits (growth at 6% p.a.) ($)	Social travel benefits (growth at 4% p.a.) ($)	Total benefits ($)	Less maintenance costs ($)	Net benefits ($)	Present worth at 5% ($)	10% ($)	15% ($)	20% ($)
1970	1,446								
1971									
1972		2,377							
1973									
1974									
1975									
1976	2,174	2,892	5,066	1,800	3,266	3,110	2,969		
1977	2,305	3,008	5,313	1,800	3,513	3,186	2,903		
1978	2,443	3,128	5,571	1,800	3,771	3,258	2,833		
1979	2,590	3,253	5,843	1,800	4,043	3,326	2,761		
1980	2,745	3,383	6,128	1,800	4,328	3,557	2,687		
1981	2,910	3,519	6,429	1,800	4,629	3,454	2,613		
1982	3,084	3,659	6,743	1,800	4,943	3,520	2,537		
1983	3,269	3,806	7,075	1,800	5,275	3,570	2,461		
1984	3,465	3,958	7,423	1,800	5,623	3,625	2,385		
1985	3,673	4,116	8,089	1,800	6,289	3,861	2,425		
						34,467	26,574		

Benefit-cost ratio at 5% = 1.15 10% = 0.89

Internal rate of return = 7.5%

Schedule 11 Siwai

	Sininai	Kumuki	Musiminoi	Miheru	Iru	Total
No of cocoa trees, 1970	-	9,641	5,030	11,812	11,933	26,483
Population 1972	-	122	113	128	152	515

Length of road: 7 miles of which 3 new

Economic benefits start at: $\$2 \times \frac{26,483}{373} \times 2 \times \frac{(18.9 \times 4 + 9.5 \times 3)}{100} + \frac{3}{10} \times \frac{26,483}{373} \times 44.80 = \$296 + 954 = \$1,250$

Social travel benefits start at: $\$ \frac{515}{25} \times 50 \times 2(1.041) = \$2,144$

Costs:

Construction: 7 miles at $5,000/mile = $35,000

Maintenance: 7 miles at $300/mile = $2,100

Year	Economic benefits (growth at 6% p.a.)	Social travel benefits (growth at 4% p.a.)	Total benefits	Less maintenance costs	Net benefits	Present worth at 5%	10%	15%	20%
	($)	($)	($)	($)	($)	($)	($)	($)	($)
1970	1,250								
1971									
1972		2,144							
1973									
1974									
1975									
1976	1,880	2,609	4,489	2,100	2,389	2,275			
1977	1,992	2,713	4,705	2,100	2,605	2,363			
1978	2,112	2,821	4,933	2,100	2,833	2,447			
1979	2,239	2,934	5,173	2,100	3,073	2,528			
1980	2,373	3,052	5,425	2,100	3,325	2,605			
1981	2,515	3,174	5,689	2,100	3,589	2,678			
1982	2,666	3,301	5,967	2,100	3,867	2,748			
1983	2,826	3,433	6,259	2,100	4,159	2,815			
1984	2,996	3,570	6,566	2,100	4,466	2,879			
1985	3,175	3,713	6,888	2,100	4,788	2,939			
						26,277			

Benefit-cost ratio at 5% = 0.75

Schedule 12 Siwai

	Nukui	Usokoli	Tohu	Total
No. of cocoa trees, 1970	4,760	6,850	1,922	13,532
Population 1972	105	131	67	303

Length of road (miles) 2½

Economic benefits start at $\$2 \times \frac{13,532}{373} \times 2 \times \frac{18.9 \times 2.5}{100} + \frac{3}{10} \times \frac{13,532}{373} \times 44.80 = \$69 + 488 = \$557$

Social travel benefits start at $\$ \frac{303}{25} \times 50 \times 2 \times \frac{18.9 \times 2.5}{100} = \$573$

Costs:

Construction 2½ miles at $5,000/mile = $12,500

Maintenance 2½ miles at $300/mile = $750

Year	Economic benefits (growth at 6% p.a.) ($)	Social travel benefits (growth at 4% p.a.) ($)	Total benefits ($)	Less maintenance costs ($)	Net benefits ($)	Present worth at 5% ($)	Present worth at 10% ($)	Present worth at 15% ($)	Present worth at 20% ($)
1970	557								
1971									
1972		573							
1973									
1974									
1975									
1976	838	697	1,535	750	785	748			
1977	888	725	1,613	750	888	805			
1978	941	754	1,695	750	945	816			
1979	998	784	1,782	750	1,032	849			
1980	1,057	816	1,873	750	1,123	880			
1981	1,121	848	1,969	750	1,219	910			
1982	1,188	882	2,070	750	1,320	938			
1983	1,259	917	2,176	750	1,426	965			
1984	1,335	954	2,289	750	1,539	992			
1985	1,415	992	2,407	750	1,657	1,017			
						8,920			

Benefit-cost ratio at 5% = 0.71

Schedule 13 Siwai

	Ununai	Mamagota	Kotu	Runai	Aitara	Total	
No. of cocoa trees, 1970	-	5,250	3,900	2,153	nil	11,303	Assume 5 tons of copra p.a. from this area. Farm gate price of copra = \$127/ton.
Population 1972	-	101	88	71	95	355	

Length of road (miles) 13.5

Economic benefits start at $\$2 \times \frac{11,303}{373} \times 2 + 13.5 \times \frac{18.9}{100} + \frac{3}{10} \times \frac{11,303}{373} \times 44.80 + 2 \times 5 \times 13.5 \times \frac{18.9}{100} + \frac{30}{100} \times 2 \times 5 \times 127 = \$309 + 407 + 26 + 381 = \$716 + 407$

Social travel benefits start at $\$ \frac{355}{25} \times 50 \times 2 \times 13.5 \times \frac{18.9}{100} = \$3,623$

Costs:

Construction 13.5 miles at \$5,000/mile = \$67,500

Maintenance 13.5 miles at \$300/mile = \$4,050

Year	Economic benefits Cocoa (\$)	Copra (\$)	Social travel benefits (growth at 4% p.a.) (\$)	Total benefits (\$)	Less maintenance costs (\$)	Net benefits (\$)	Present worth at 5% (\$)	10% (\$)	15% (\$)	20% (\$)
1970	716	407								
1971		407								
1972		407	3,623							
1973		407								
1974		407								
1975		407								
1976	1,077	407	4,408	5,892	4,050	1,842	1,754			
1977	1,141	407	4,584	6,132	4,050	2,082	1,888			
1978	1,210	407	4,768	6,385	4,050	2,335	2,017			
1979	1,282	407	4,958	6,647	4,050	2,597	2,137			
1980	1,359	407	5,157	6,923	4,050	2,873	2,251			
1981	1,441	407	5,363	7,211	4,050	3,161	2,359			
1982	1,527	407	5,577	7,511	4,050	3,461	2,460			
1983	1,619	407	5,801	7,827	4,050	3,777	2,556			
1984	1,716	407	6,033	8,156	4,050	4,106	2,647			
1985	1,819	407	6,274	8,500	4,050	4,450	2,732			
							22,801			

Benefit-cost ratio at 5% = 0.34

Schedule 14 Buin

	Mamaramino (1)	Mamaramino (2)	Mamaramino (3)	Luaguo	Kogu	Muguai	Taguruai	Iamaru	Lakoei	Kiki-batsiogu	Komai	Ka'amoro	Tubaro	Tabago	Moriu	Ipilai	Siul	Laitaro	Okoiragu	Kukumai	Total
No. of cocoa trees, 1970	14,939	2,597	9,185	4,075	10,532	22,046	8,786	6,896	5,551	12,173	2,907	15,047	5,583	5,159	5,948	9,909	5,270	13,381	4,878	12,748	177,610
Population 1972	188		150	124	78	151	83	83	107	59	20	196	86	119	72	190	98	271	117	116	2,308

Length of road: 20 miles of which 12 new

Economic benefits start at: $\$2 \times \frac{177,610}{373} \times 2 \times (\frac{18.9 \times 8 + 9.5 \times 12}{100}) + (\frac{3}{10} \times \frac{177,610}{373} \times 44.80) = \$5,051 + 6,400 = \$11,451$

Social travel benefits start at: $\$\ \frac{2,308}{25} \times 50 \times 2(\frac{18.9 \times 8 + 9.5 \times 12}{100}) = \$24,483$

Costs:

Construction: 20 miles at \$5,000/mile = \$100,000

Maintenance: 20 miles at \$300/mile = \$6,000

Year	Economic benefits (growth at 6% p.a.) ($)	Social travel benefits (growth at 4% p.a.) ($)	Total benefits ($)	Less maintenance costs ($)	Net benefits ($)	Present worth at 5% ($)	10% ($)	15% ($)	20% ($)
1972		24,483							
1973	11,451								
1974									
1975									
1976	14,457	29,787	44,244	6,000	38,244	36,423	34,767	33,256	31,870
1977	15,324	30,979	46,303	6,000	40,303	36,556	33,308	30,475	27,988
1978	16,243	32,218	48,461	6,000	42,461	36,679	31,902	27,919	24,572
1979	17,218	33,507	50,725	6,000	44,725	36,795	30,548	25,572	21,569
1980	18,251	34,847	53,098	6,000	47,098	36,903	29,244	23,416	18,928
1981	19,346	36,241	55,587	6,000	49,587	37,003	27,991	21,438	16,607
1982	20,507	37,690	58,197	6,000	52,197	37,095	26,785	19,623	14,567
1983	21,737	39,198	60,935	6,000	54,935	37,182	25,628	17,958	12,776
1984	23,042	40,766	63,808	6,000	57,808	37,264	24,516	16,433	11,204
1985	24,424	42,397	66,821	6,000	60,821	37,339	23,449	15,034	9,823
						369,239	288,138	230,124	189,905

Benefit-cost ratio at 5% = 3.7 10% = 2.9 15% = 2.3 20% = 1.9

Internal rate of return = 41%

<u>Schedule 15</u> Buin

	Tabago	Pililalo	Porero	Kogwiriru	Mogoroi	Ororoi	Orumoi	Pauroko	Koniguru	Kumirogu	Kugugai	Pariro	Iula (1)	Iula (2)	Total
No. of cocoa trees, 1973	-	13,110	5,698	4,426	21,883	9,348	18,375	26,465	7,507	3,501	11,537	47,465	18,210	-	187,525
Population 1972	-	100	60	74	148	103	73	116	198	76	91	378	129	-	1,546

Length of road: Old 1.5 miles; New 13 miles

Economic benefits start at $\$2 \times \frac{187,525}{373} \times 2 \times (\frac{18.9 \times 1.5 + 9.5 \times 13}{100}) + (\frac{3}{10} \times \frac{187,525}{373} \times 44.80) = \$3,054 + 6,757 = \$9,811$

Social travel benefits start at $\$ \frac{1,546}{25} \times 50 \times 2(\frac{18.9 \times 1.5 + 9.5 \times 13}{100}) = \$1,546 \times 2 \times 2(1.5185) = \$9,390$

Costs:

Construction: 14½ miles at $5,000/mile = $72,500
Maintenance: 14½ miles at $300/mile = $4,350

Year	Economic benefits (growth at 6% p.a.) ($)	Social travel benefits (growth at 4% p.a.) ($)	Total benefits ($)	Less maintenance costs ($)	Net benefits ($)	Present worth at 5% ($)	10% ($)	15% ($)	20% ($)	30% ($)
1972	9,811	9,390								
1973										
1974										
1975										
1976	12,386	11,424	23,810	4,350	19,460	18,533	17,691	16,922	16,217	14,969
1977	13,129	11,881	25,010	4,350	20,660	18,739	17,074	15,622	14,347	12,225
1978	13,917	12,357	26,274	4,350	21,924	18,939	16,472	14,415	12,688	9,979
1979	14,752	12,851	27,603	4,350	23,253	19,130	15,882	13,295	12,214	8,142
1980	15,637	13,365	29,002	4,350	24,652	19,315	15,307	12,256	9,907	6,640
1981	16,575	13,899	30,474	4,350	26,124	19,494	14,746	11,294	8,749	5,412
1982	17,570	14,455	32,025	4,350	27,675	19,668	14,202	10,404	7,724	4,410
1983	18,624	15,034	33,658	4,350	29,308	19,837	13,672	9,581	6,816	3,593
1984	19,742	15,635	35,377	4,350	31,027	20,000	13,158	8,820	6,013	2,926
1985	20,926	16,260	37,186	4,350	32,836	20,158	12,660	8,117	5,303	2,382
						193,813	150,864	120,726	98,978	70,678

Benefit-cost ratio at 5% = 2.7 10% = 2.1 15% = 1.7 20% = 1.4

Internal rate of return = 29%

Schedule 16 Buin

	Kikimogu	Mouakei	Ibiro	Tugiogu	Toaragai	Tugiu	Turigau	Tsimbo	Lukaruru	Morula	Kogisagano	Lobigau	Omitaro	Numakei	Parerono	Total
No. of cocoa trees, 1973	-	5,933	10,735	24,736	9,931	10,768	9,756	8,327	12,512	7,485	2,802	3,455	4,299	7,202	16,816	134,757
Population 1972	-	83	73	311	98	260	132	116	257	110	89	71	57	93	152	1,902

Length of road (miles) 23

Economic benefits start at $\$2 \times \frac{134,757}{373} \times 2 \times 23 \times \frac{9.5}{100} + \frac{3}{10} \times \frac{134,757}{373} \times 44.80 = \$3,158 + 4,856 = \$8,014$

Social travel benefits start at $\$ \frac{1,902}{25} \times 50 \times 2 \times 2.3 \times \frac{9.5}{100} = \$16,623$

Costs:

Construction 23 miles at $5,000/mile = $115,000

Maintenance 23 miles at $300/mile = $6,900

Year	Economic benefits (growth at 6% p.a.) ($)	Social travel benefits (growth at 4% p.a.) ($)	Total benefits ($)	Less maintenance costs ($)	Net benefits ($)	Present worth at 5% ($)	10% ($)	15% ($)	20% ($)
1972		16,623							
1973	8,014								
1974									
1975									
1976	10,117	20,224	30,341	6,900	23,441	22,325	21,310	20,383	19,534
1977	10,725	21,033	31,758	6,900	24,858	22,547	20,544	18,796	17,263
1978	11,368	21,875	33,243	6,900	26,343	22,756	19,792	17,321	15,245
1979	12,050	22,750	34,800	6,900	27,900	22,953	19,056	15,952	13,455
1980	12,773	23,660	36,433	6,900	29,533	23,140	18,338	14,683	11,869
1981	13,539	24,606	38,145	6,900	31,245	23,316	17,637	13,508	10,464
1982	14,352	25,590	39,942	6,900	33,042	23,482	16,956	12,422	9,221
1983	15,213	26,614	41,827	6,900	34,927	23,640	16,294	11,418	8,123
1984	16,126	27,679	43,805	6,900	36,905	23,789	15,651	10,495	7,152
1985	17,093	28,786	46,689	6,900	39,789	24,427	15,340	9,835	6,426
						232,375	181,918	144,809	118,752

Benefit-cost ratio at 5% = 2.0 10% = 1.6 15% = 1.3 20% = 1.03

Internal rate of return = 20.5%

Schedule 17 Buin

	Luilau	Kaitu	Lukauku	Bogisago	Total
No. of cocoa trees, 1973	-	22,744	24,989	33,319	81,052
Population 1972	-	96	190	228	514
Length of road (miles)	4				

Economic benefits start at $\$2 \times \frac{81,052}{373} \times 2 \times 4 \times \frac{9.5}{100} + \frac{3}{10} \times \frac{81,052}{373} \times 44.80 = \$3,250$

Social travel benefits start at $\$ \frac{514}{25} \times 50 \times 2 \times 4 \times \frac{9.5}{100} = \$781$

Costs:

Construction 4 miles at $5,000/mile = $20,000

Maintenance 4 miles at $300/mile = $1,200

Year	Economic benefits (growth at 6% p.a.) ($)	Social travel benefits (growth at 4% p.a.) ($)	Total benefits ($)	Less maintenance costs ($)	Net benefits ($)	Present worth at 5% ($)	10% ($)	15% ($)	20% ($)
1972		781							
1973	3,250								
1974									
1975									
1976	4,103	950	5,053	1,200	3,853	3,670	3,503	3,350	3,211
1977	4,349	988	5,337	1,200	4,137	3,752	3,419	3,128	2,873
1978	4,610	1,028	5,638	1,200	4,438	3,834	3,334	2,918	2,568
1979	4,887	1,069	5,956	1,200	4,756	3,913	3,248	2,719	2,294
1980	5,180	1,117	6,297	1,200	5,097	3,994	3,165	2,534	2,048
1981	5,491	1,156	6,647	1,200	5,447	4,065	3,075	2,355	1,824
1982	5,820	1,202	7,022	1,200	5,822	4,138	2,988	2,189	1,625
1983	6,169	1,250	7,419	1,200	6,219	4,209	2,901	2,033	1,446
1984	6,540	1,300	7,840	1,200	6,640	4,280	2,816	1,888	1,287
1985	6,932	1,352	8,284	1,200	7,084	4,349	2,731	1,751	1,144
						40,204	31,180	24,864	20,320

Benefit-cost ratio at 5% = 2.01 10% = 1.56 15% = 1.24 20% = 1.02

Internal rate of return = 20.4%

Schedule 18 Buin

	Buin	Maramoku	Moro	Kanauro	Kikimogu	Piarino	Iula (2)	Iula (3)	Kukumaru	Laitaro	Total
No. of cocoa trees, 1973	-	348	20,648	23,007	15,509	5,115	11,918	10,811	7,441	-	94,797
Population 1972	-	107	180	255	148	78	117	83	256	-	1,224

Length of road: 15.5 miles of which 10 new

Economic benefits start at: $\$2 \times \frac{94,797}{373} \times 2(\frac{18.9 \times 5.5 + 9.5 \times 10}{100}) + (\frac{3}{10} \times \frac{94,797}{373} \times 44.80) = \$2,023 + 3,416 = \$5,439$

Social travel benefits start at: $\$ \frac{1,224}{25} \times 50 \times 2(1.9895) = \$9,741$

Costs:

Construction: 15.5 miles at $5,000/mile = $77,500

Maintenance: 15.5 miles at $300/mile = $4,650

Year	Economic benefits (growth at 6% p.a.) ($)	Social travel benefits (growth at 4% p.a.) ($)	Total benefits ($)	Less maintenance costs ($)	Net benefits ($)	Present worth at 5% ($)	10% ($)	15% ($)	20% ($)
1972		9,741							
1973	5,439								
1974									
1975									
1976	6,867	11,851	18,718	4,650	14,068	13,398	12,789	12,233	11,723
1977	7,279	12,325	19,604	4,650	14,954	13,564	12,359	11,307	10,385
1978	7,715	12,818	20,533	4,650	15,883	13,720	11,933	10,443	9,192
1979	8,178	13,331	21,509	4,650	16,859	13,870	11,515	9,639	8,130
1980	8,669	13,864	22,533	4,650	17,883	14,012	11,104	8,891	7,187
1981	9,189	14,419	23,608	4,650	18,958	14,148	10,701	8,196	6,349
1982	9,740	14,996	24,736	4,650	20,086	14,275	10,307	7,551	5,606
1983	10,325	15,596	25,921	4,650	21,271	14,397	9,923	6,954	4,947
1984	10,944	16,219	27,163	4,650	22,513	14,512	9,548	6,400	4,363
1985	11,601	16,868	28,469	4,650	23,819	14,623	9,183	5,888	3,847
						140,519	109,362	87,232	71,731

Benefit-cost ratio at 5% = 1.81 10% = 1.41 15% = 1.13 20% = 0.93

Internal rate of return = 17.5%

Schedule 19 Buin

	Tabago	Luilau	Turitai	Ububakogu	Oria	Pagui	Orimai	Botulai	Sulukunu	Kaukausina	Kekemona	Total
No. of cocoa trees, 1973	-	9,069	10,039	4,110	21,244	21,870	21,128	9,054	8,490	1,736	765	107,505
Population 1972	-	88	64	61	157	195	273	106	189	83	102	1,318

Length of road (miles) 20

Economic benefits start at $\$2 \times \frac{107,505}{373} \times 2 \times 20 \times \frac{9.5}{100} + \frac{3}{10} \times \frac{107,505}{373} \times 44.80 = \$2,190 + 3,874 = \$6,064$

Social travel benefits start at $\$ \frac{1,318}{25} \times 50 \times 2 \times 20 \times \frac{9.5}{100} = \$10,017$

Costs:

Construction 20 miles at \$5,000/mile = \$100,000

Maintenance 20 miles at \$300/mile = \$6,000

Year	Economic benefits (growth at 6% p.a.)	Social travel benefits (growth at 4% p.a.)	Total benefits	Less maintenance costs	Net benefits	Present worth at			
						5%	10%	15%	20%
	(\$)	(\$)	(\$)	(\$)	(\$)	(\$)	(\$)	(\$)	(\$)
1972		10,017							
1973	6,064								
1974									
1975									
1976	7,657	12,187	19,844	6,000	13,844	13,185	12,585	12,038	
1977	8,115	12,675	20,790	6,000	14,790	13,415	12,223	11,183	
1978	8,602	13,181	21,783	6,000	15,783	13,634	11,858	10,378	
1979	9,118	13,709	22,827	6,000	16,827	13,844	11,493	9,621	
1980	9,665	14,257	23,922	6,000	17,922	14,042	11,128	8,910	
1981	10,245	14,828	25,073	6,000	19,073	14,233	10,766	8,246	
1982	10,860	15,421	26,281	6,000	20,281	14,413	10,407	7,624	
1983	11,511	16,038	27,549	6,000	21,549	14,585	10,053	7,044	
1984	12,202	16,679	28,881	6,000	22,881	14,749	9,704	6,504	
1985	12,934	17,346	30,280	6,000	24,280	14,906	9,360	6,002	
						141,006	109,577	87,550	

Benefit-cost ratio at 5% = 1.41 10% = 1.10 15% = 0.88

Internal rate of return = 11.8%

Schedule 20 Buin

	Omitaro	Borubi	Okomo	Pamaiuta	Buraburunna	Togulegu	Matsiogu	Total
No. of cocoa trees, 1973	-	3,232	1,433	14,230	1,032	n.a.	2,253	22,270
Population 1972	-	79	94	111	110	37	106	537

Length of road (miles) 7.5

Economic benefits start at $\$2 \times \frac{22,270}{373} \times 2(\frac{9.5 \times 7.5}{100}) + \frac{3}{10} \times (\frac{22,270}{373} \times 44.80) = \$170 + 802 = \$972$

Social travel benefits start at $\$ \frac{537}{25} \times 53 \times 2 \times \frac{9.5 \times 7.5}{100} = \$1,530$

Costs:

Construction 7.5 miles at $5,000/mile = $37,500

Maintenance 7.5 miles at $300/mile = $2,250

Year	Economic benefits (growth at 6% p.a.) ($)	Social travel benefits (growth at 4% p.a.) ($)	Total benefits ($)	Less maintenance costs ($)	Net benefits ($)	Present worth at 5% ($)	Present worth at 10% ($)	Present worth at 15% ($)	Present worth at 20% ($)
1972		1,530							
1973	972								
1974									
1975									
1976	1,227	1,861	3,088	2,250	838	798			
1977	1,301	1,936	3,237	2,250	987	895			
1978	1,379	2,013	3,392	2,250	1,142	987			
1979	1,462	2,094	3,556	2,250	1,306	1,074			
1980	1,549	2,178	3,727	2,250	1,477	1,157			
1981	1,642	2,265	3,907	2,250	1,657	1,236			
1982	1,741	2,355	4,096	2,250	1,846	1,312			
1983	1,845	2,450	4,295	2,250	2,045	1,384			
1984	1,956	2,548	4,504	2,250	2,254	1,453			
1985	2,073	2,649	4,722	2,250	2,472	1,518			
					16,024	10,430			

Benefit-cost ratio at 5% = 0.28%

Schedule 21 Buin

	Lukaruru	Tubaro	Tubobisou	Siuru	Piruruino	Total
No. of cocoa trees, 1973	-	19,197	4,148	8,823	3,532	35,700
Population 1972	-	165	92	122	61	440
Length of road (miles)	7					

Economic benefits start at $\$2 \times \frac{35,700}{373} \times 2 \times \frac{9.5 \times 7}{100} + \frac{3}{10} \times \frac{35,700}{373} \times 44.80 = \$254 + 1,286 = \$1,540$

Social travel benefits start at $\$ \frac{440}{25} \times 50 \times 2 \times \frac{9.5 \times 7}{100} = \$1,170$

Costs:

Construction 7 miles at $5,000/mile = $35,000

Maintenance 7 miles at $300/mile = $2,100

Year	Economic benefits (growth at 6% p.a.) ($)	Social travel benefits (growth at 4% p.a.) ($)	Total benefits ($)	Less maintenance costs ($)	Net benefits ($)	Present worth at 5% ($)	10% ($)	15% ($)	20% ($)
1972		1,170							
1973	1,540								
1974									
1975									
1976	1,944	1,423	3,367	2,100	1,267	1,207			
1977	2,061	1,480	3,541	2,100	1,441	1,307			
1978	2,185	1,540	3,725	2,100	1,625	1,404			
1979	2,316	1,601	3,917	2,100	1,817	1,495			
1980	2,455	1,665	4,120	2,100	2,020	1,583			
1981	2,602	1,732	4,334	2,100	2,234	1,667			
1982	2,758	1,801	4,559	2,100	2,459	1,748			
1983	2,923	1,873	4,796	2,100	2,696	1,825			
1984	3,099	1,948	5,047	2,100	2,947	1,900			
1985	3,285	2,026	5,311	2,100	3,211	1,971			
					21,717	16,107			

Benefit-cost ratio at 5% = 0.46

Schedule 22 Buin

	Kogu	Maika	Tandareki	Togurau Plantation	Moisuru	Total
No. of cocoa trees, 1973	-	10,895	8,407	n.a.	nil	19,302
Population 1972	-	105	189	25 (est.)	73	392

Assume 5 tons copra p.a. from Moisuru. Assume 15 tons copra and 15 tons cocoa p.a. from Togurau Plantation.

Length of road (miles) 5

Economic benefits start at $\$2 \times \frac{19,302}{373} \times 2 \times 5 \times \frac{18.9}{100} + \frac{3}{10} \times \frac{19,302}{373} \times 44.80 + 2 \times 15 \times 10 \times \frac{18.9}{100} + \frac{30}{100} \times 2 \times 15 \times 127 + 2 \times 30 \times 2 \times 5 \times \frac{18.9}{100} = \$196 + 695 + 381 = \$891 + 381$

Social travel benefits start at $\$ \frac{392}{25} \times 50 \times 2 \times 5 \times \frac{18.9}{100} = \$1,482$

Costs:

Construction 5 miles at \$5,000/mile = \$ 25,000

Maintenance 5 miles at \$300/mile = \$1,500 p.a.

Year	Economic benefits Cocoa ($)	Economic benefits Copra ($)	Social travel benefits (growth at 4% p.a.) ($)	Total benefits ($)	Less maintenance costs ($)	Net benefits ($)	Present worth at 5% ($)	Present worth at 10% ($)	Present worth at 15% ($)	Present worth at 20% ($)
1972			1,482							
1973	891	381								
1974		381								
1975		381								
1976	1,125	381	1,803	3,309	1,500	1,809	1,723			
1977	1,192	381	1,875	3,448	1,500	1,948	1,767			
1978	1,264	381	1,950	3,595	1,500	2,095	1,810			
1979	1,340	381	2,028	3,749	1,500	2,249	1,850			
1980	1,420	381	2,109	3,910	1,500	2,410	1,888			
1981	1,505	381	2,194	4,080	1,500	2,580	1,925			
1982	1,596	381	2,281	4,258	1,500	2,758	1,960			
1983	1,691	381	2,373	4,445	1,500	2,945	1,993			
1984	1,793	381	2,468	4,642	1,500	3,142	2,025			
1985	1,900	381	2,566	4,847	1,500	3,347	2,055			
							18,996			

Benefit-cost ratio at 5% = 0.76

Appendix VI

Sources of maps

1.1	Department of District Administration: District map of Southern Bougainville
2.1	CSIRO, 1967:8
2.2	CSIRO, 1967:Map sheet
2.3	CSIRO, 1967:85
2.4	CSIRO, 1967:Table 1, p.63; Table 4, p.67; Brookfield and Hart, 1966:Map 5; BCP, Panguna
2.5-2.8	CSIRO, 1967:Map sheet
2.9	Papua New Guinea, Department of District Administration, Village Population Registers, 1972
2.10	Allen, J. and Hurd, C., 1963
2.11	Brown, M.F.K., 1972:735
3.1	Papua New Guinea, Department of Transport, File 68/73(3); DASF, Buin, 1972
3.2	Papua New Guinea Statistical Bulletin, Overseas Trade, Nov.-Dec. 1971
3.3	Boku Patrol Report No.3 of 1970-71
3.4	Konga Patrol Report No.1 of 1970-71; DASF, Buin, 18 Oct. 1973; Boku Patrol Report No.3 of 1970-71
3.5	Papua New Guinea, DASF, Buin; Smith, 1972:11; Co-operatives Development Officer, Buin, Annual Report, 1972; P.N.G. Department of Transport File 68/73(3)
3.6	Siwai Co-operative Producers' Society
3.7	P.N.G. Department of Business Development, Co-operatives Division, Port Moresby
3.8 and 3.9	BCP
3.10	Local Government Council Records, Bana, Siwai and Buin
3.11	Development Bank, Port Moresby
4.1	DDA, Village censuses
4.2	BCP, Labour records
5.1	Department of Lands, Surveys and Mines, Port Moresby
6.1	Australian Survey Corps, c.1944, 1:63,360

Nos.4895	Torokina	Nos.4894	Puriata River
3291	Kahili	3327	Mt Taroka
3328	Toimonapu	3159	Aitara
3120	Kieta	3254	Kara

6.2	Papua New Guinea: Bureau of Statistics, Papua New Guinea Statistical Bulletins, Census of Motor Vehicles Registered at 31 December, 1967-71
6.3	Vehicle Registration Duplicates, Subdistrict Office, Buin
6.4	Duplicates of Vehicle Registrations, Buin town
6.5	Duplicates of PMV licences issued at Buin
6.6 and 6.7	Papua New Guinea: Bureau of Statistics, _Statistical Bulletin_: Licences issued to drivers and riders of motor vehicles, years ended 31 December 1967, 1969 and 1971
6.8 and 6.9	Smith, D.N., 1972; Public Works Department, PNG, f.p.18.
7.1	DCA, Port Moresby
8.1, 8.2 and 8.3	Department of District Administration, District Map of Southern Bougainville
AV.1	Calculated

Bibliography

Official sources

Papua New Guinea, Bougainville District Annual Report, 1964-65. Typescript.

-- Bougainville District Annual Report, 1969-70.

-- Bureau of Statistics: Population Census, July 1971. Preliminary. Bulletin no.1, Urban Centres. Port Moresby, 1972.

-- Bureau of Statistics: Statistical Bulletin: licences issued to drivers and riders of motor vehicles. Years ended 31 December 1967, 1969, 1971. Port Moresby.

-- Bureau of Statistics: Statistical Bulletin, 20 December 1973. Port Moresby.

-- Central Planning Office: Papua New Guinea's Improvement Plan 1973-74. Port Moresby, 1973 and Districts Supplement, 1973.

-- Co-operatives Development Officer, Buin, Annual Report, 1971/72.

-- Department of Agriculture, Stock and Fisheries: Fresh Food Supplies for the Towns of Kieta, Arawa and Panguna. 1971.

-- Department of Business Development, Co-operatives Division: card file, Port Moresby.

-- Department of District Administration: Village Directory. Port Moresby, 1973.

-- Department of Trade and Industry, Office of the Economic Adviser (c.1969): Survey of Indigenous-owned Trade Stores 1968-1969. Port Moresby.

-- Department of Trade, Co-operatives Division, File BD 12-4-1, Co-operatives Monthly Reports, Buin.

-- Department of Transport: 1968, 1969, 1970. Highlands Highway Freight Survey. Computer output. Port Moresby.

-- Department of Transport: 'Report on the history and validity of proposals for a road, Panguna to Lambalam/Beretemba, South Bougainville'. Typescript 1970.

-- Department of Transport: File 68/73(3) Bougainville District Roads General, Port Moresby, 1972.

-- Patrol Reports: Konga, 1970-71; Boku, 1970-71; Bougainville, 1946-49.

-- Village Population Registers for Banoni, Nagovisi, Baitsi, Siwai, Makis, Lugakei, Paubakei and Kono census divisions, 1972.

Territory of Papua New Guinea, Bureau of Statistics: Survey of indigenous agriculture and ancillary surveys 1961-62. Port Moresby, 1963.

Non-official sources

Allen, J. and Hurd, C., 1963. 'Languages of the Bougainville District'. Summer Institute of Linguistics, mimeograph.

Bedford, R.D., 1974. 'New towns in Papua New Guinea: Bougainville'. Australian Geographer, 12, 6, 543-53.

Bonney, R.S.P., 1964. The relationship between road building and economic and social development in Sabah; Road Research Laboratory, Department of Scientific and Industrial Research; Crowthorne (England).
Part 1: Roads and traffic (note LN 519).
Part 2: Development and the use of other transport in relation to road transport (note LN 520).
Part 3: Agricultural development, traffic and road standards (note LN 648).

Bouchard, J.F., 1972. The impact of roads on the monetary activities of subsistence economies, in the Okapa region of Papua New Guinea. University of Papua and New Guinea, Department of Geography, Occasional Paper No.4. Port Moresby.

-- 1973. The impact of roads in the Gumine Region of Papua New Guinea. University of Papua New Guinea, Department of Geography, Occasional Paper No.5. Port Moresby.

Brookfield, H.C. and Hart, Doreen, 1966. Rainfall in the Tropical Southwest Pacific. Research School of Pacific Studies, Department of Geography Publication G3 (1960), A.N.U., Canberra.

Brown, M.J.F., 1972. 'Bougainville pays in copper'. The Geographical Magazine, vol.XLIV, no.11, pp.734-6.

-- 1973. 'A development consequence - disposal of mining waste on Bougainville', ANZAAS paper, Perth, mimeo.

Commonwealth Scientific and Industrial Research Organisation, 1967. Lands of Bougainville and Buka Islands, Territory of Papua and New Guinea. Papers by R.M. Scott, P.B. Heyligers, J.R. McAlpine, J.C. Saunders and J.G. Speight. Melbourne.

Couper, A.D. and Cotter, C.H., 1972. Bougainville. A Report on Transport and Development in Bougainville Island, Territory of Papua New Guinea. University of Wales, Institute of Technology, Cardiff.

Emmery, P., 1970. 'Increasing economic efficiency from plantation tree crops'. ANZAAS 42nd Congress, mimeo.

Griffin, J., 1972. 'Bougainville - secession or just sentiment', Current Affairs Bulletin, vol.48, no.9. Sydney.

Halcrow, Sir William, and Partners, 1969. Transport survey of the Territories of Papua and New Guinea, 6 vols, London, for International Bank for Reconstruction and Development, United Nations Development Programme.

Heyligers, P.C., 1967. 'Vegetation and Ecology of Bougainville and Buka Islands', pp.121-45 in CSIRO, Lands of Bougainville and Buka Islands, Territory of Papua and New Guinea. Melbourne.

Jones, J.H., 1964. Economic benefits from development roads in Thailand, SEATO Graduate School of Engineering, Bangkok, Technical note no.15.

Jones, J.H. and Orr, J.A., 1966. Exploratory study of economic effects of selected village roads in northeast Thailand, SEATO Graduate School of Engineering, Bangkok, Technical note no.25.

Kasiraksa, W., 1963. 'Economic effects of the Friendship Highway'. Unpublished master's thesis, SEATO Graduate School of Engineering, Bangkok.

Laracy, H.M., 1969. 'Catholic Missions in the Solomon Islands, 1845-1966'. Unpublished Ph.D. thesis, A.N.U., Canberra.

Long, G., 1963. Australia in the War of 1939-1945, vol.VII, The Final Campaigns, Canberra, 1963.

Mamak, A. and Bedford, R., with Leo Hannett and Moses Havini, 1974. Bougainvillean nationalism, aspects of unity and discord. Bougainville Special Publication no.1. Christchurch.

Mamak, A. and Bedford, R.D., 1974. 'Bougainville's students: some expressed feelings towards non-Bougainvilleans, Arawa town, and the coppermining company', New Guinea and Australia, the Pacific and South-East Asia, 9, 1.

McAlpine, J.R., 1967. 'Population and Land Use of Bougainville and Buka Islands' in CSIRO, Lands of Bougainville and Buka Islands, Territory of Papua and New Guinea. Melbourne.

Mitchell, D.D., 1971. 'Gardening for money: land and agriculture in Nagovisi'. Unpublished Ph.D. thesis, Harvard University.

Mitchell, Jill Nash, 1971. 'Aspects of matriliny in Nagovisi society'. Unpublished Ph.D. thesis, Harvart University. Published as 'Matriliny and modernisation: the Nagovisi of South Bougainville', New Guinea Research Bulletin, no.55, 1974.

Moulik, T.K., 1974 'Bougainville nationalism: problems and prospects'. Mimeo, 20 pp.

Mount, K., 1973. 'Land available in PNG for the expansion of commercial agriculture and the population absorptive capacity of the land'. Typescript.

Ngamsit, S., 1966. Study conducted for United States' Operations Mission to Thailand (USOM), unpublished report. Mimeo.

Ogan, E., 1970. 'The Nasioi vote again', Human Organisation, vol.29, no.3, pp.178-89.

-- 1971. 'Charisma and race' in The Politics of Dependence: Papua New Guinea 1968. A.L. Epstein, R.S. Parker and M. Reay, eds. Canberra.

-- 1972. 'Business and Cargo', New Guinea Research Bulletin, no.44, Canberra.

Oliver, D.L., 1955. A Solomon Island Society: Kinship and Leadership among the Siwai of Bougainville. Harvard University Press, Cambirdge.

Patanapanich, T., 1964. Economic effects of the East-West Highway. SEATO Graduate School of Engineering, Bangkok. Thesis no.62.

Philpott, M., 1972. Economic Development in the Sepik River Basin. An appraisal of transport needs and potential in 1971-72. Joint Programme of Studies in the Transport Process. Port Moresby.

Rimmer, P.J., 1972. 'Freight Forwarding in the Australia-Papua New Guinea Trade', New Guinea Research Bulletin, no.48. Canberra.

Sabag, P.L., 1971. 'Some impacts of new road construction: an example from Sabah'. Unpublished thesis submitted to the James Cook University of North Queensland as part of the requirement for the degree of Bachelor of Arts with Honours (Geography).

Scott, R.M., 1967. 'Land use capability of Bougainville and Buka Islands' in CSIRO, Lands of Bougainville and Buka Islands, Territory of Papua and New Guinea. Melbourne.

Scott, W.D. and Co. Pty Ltd, 1973. A Study of the Impact of the Bougainville Copper Project on the Economy and Society of Papua New Guinea. Sydney.

Smith, D.N., 1972. Roads needs - Buin Subdistrict - 1972, South West Bougainville, Preliminary Report, Public Works Department, Papua New Guinea. Mimeo.

Smith, N.D.S., 1959. 'A Pilot Study in Uganda of the effects upon economic development of the construction of feeder roads'. Research Note no. RN/3408/NDSS, Department of Scientific and Industrial Research, Road Research Laboratory, Harmondsworth, England. Mimeo.

Southern, R.J., 1970a. 'Local roads - Mt Hagen area', Directorate of Transport research report no.1. Mimeo.

Southern, R.J., 1970b. 'Transport and development in the Mt Hagen area', paper given at New Guinea Research Unit Seminar. Mimeo.

-- 1971. 'The transport system of the New Guinea Highlands: some implications for rural development' in Ward, Marion W. (ed.): Change and Development in Rural Melanesia, Papers of the Fifth Waigani Seminar. Canberra.

-- 1973. Road Transport in the New Guinea Highlands, University of Papua New Guinea, Department of Geography, Occasional Paper no.6, Port Moresby.

Sumbak, J.H., 1970. 'Poor coconut growth in Southwest Bougainville', Papua New Guinea Agricultural Journal, vol.22, no.1, pp.1-5. Port Moresby.

Thailand National Statistics Office, 1966, unpublished report.

Trikamol, W., 1966. Study conducted for USOM, unpublished report.

United States Operations Mission, 1966. Joint Thai-USOM evaluation of the ARD program, unpublished report.

Ward, Marion W., 1970. 'The Rigo road: a study of the economic effects of new road construction', New Guinea Research Bulletin, no.33. Canberra.

-- 1973. 'Intercensal urban growth in Papua New Guinea, 1966-71', The Australian Geographer, vol.XII, no.3, pp.271-4.

Ward, R.G., 1971. 'Internal migration and urbanisation in Papua New Guinea', New Guinea Research Bulletin, no.42, pp.81-107. Canberra.

Ward, R.G., Nancy Clark, Diana Howlett, C.C. Kissling, H.C. Weinand, 1974. Growth Centres and Area Improvement in the Eastern Highlands District, Department of Human Geography, Research School of Pacific Studies, A.N.U., Canberra.

Wilson, G.W., Barbara R. Bergman, L.V. Hirsch and M.S. Klein, 1966. The Impact of Highway Investment on Development. The Brookings Institution, Transport Research Program, Washington D.C.

Index

New Guinea Research Bulletins

Bulletin No. 1	The Erap Mechanical Farming Project by R.G. Crocombe and G.R. Hogbin, April 1963
Bulletin No. 2	Land, Work and Productivity at Inonda by R.G. Crocombe and G.R. Hogbin, August 1963
Bulletin No. 3	Social Accounts of the Monetary Sector of the Territory of Papua and New Guinea, 1956/57 to 1960/61 by R.C. White, January 1964
Bulletin No. 4	Communal Cash Cropping among the Orokaiva by R.G. Crocombe, May 1964
Bulletin No. 5	A Survey of Indigenous Rubber Producers in the Kerema Bay Area by G.R. Hogbin, October 1964
Bulletin No. 6	The European Land Settlement Scheme at Popondetta by D.R. Howlett, April 1965
Bulletin No. 7	The M'buke Co-operative Plantation by R.G. Crocombe, August 1965
Bulletin No. 8	Cattle, Coffee and Land among the Wain by Graham Jackson, December 1965
Bulletin No. 9	An Integrated Approach to Nutrition and Society: the case of the Chimbu, ed. E. Hipsley, January 1966
Bulletin No.10	The Silanga Resettlement Project by Olga van Rijswijck, February 1966
Bulletin No.11	Land Tenure and Land Use among the Mount Lamington Orokaiva by Max Rimoldi, April 1966
Bulletin No.12	Education Through the Eyes of an Indigenous Urban Elite by Karol van der Veur and Penelope Richardson, August 1966
Bulletin No.13	Orokaiva Papers: Miscellaneous Papers on the Orokaiva of North East Papua, November 1966
Bulletin No.14	Rabia Camp: a Port Moresby Migrant Settlement by Nancy E. Hitchcock and N.D. Oram, January 1967
Bulletin No.15	Bulolo: a History of the Development of the Bulolo Region, New Guinea by Allan Healy, February 1967
Bulletin No.16	Papuan Entrepreneurs: Papers by R.G. Crocombe, W.J. Oostermeyer and Joanne Gray, J.V. Langmore, April 1967
Bulletin No.17	Land Tenure Conversion in the Northern District of Papua by David Morawetz, May 1967
Bulletin No.18	Social and Economic Relationships in a Port Moresby Canoe Settlement by N.D. Oram, July 1967
Bulletin No.19	A Benefit Cost Analysis of Resettlement in the Gazelle Peninsula by S. Singh, September 1967
Bulletin No.20	New Guinea People in Business and Industry: Papers from the First Waigani Seminar, December 1967

Bulletin No.21	Teachers in the Urban Community by Penelope Richardson and Karol van der Veur, January 1968
Bulletin No.22	Papers on the Papua-New Guinea House of Assembly by Norman Meller, January 1968
Bulletin No.23	Mixed-race Society in Port Moresby by B.G. Burton-Bradley, March 1968
Bulletin No.24	The Organisation of Production and Distribution among the Orokaiva by E.W. Waddell and P.A. Krinks, September 1968
Bulletin No.25	A Survey of Village Industries in Papua-New Guinea by R. Kent Wilson, November 1968
Bulletin No.26	The Contribution of Voluntary Aid Organisations to the Development of Papua-New Guinea, 1966-67, by Micheline Dewdney, January 1969
Bulletin No.27	New Guinean Entrepreneurs by B.R. Finney, February 1969
Bulletin No.28	Namasu: New Guinea's Largest Indigenous-owned Company by I.J. Fairbairn, March 1969
Bulletin No.29	Hohola: the Significance of Social Networks in Urban Adaptation of Women by Lynn Oeser, June 1969
Bulletin No.30	Inter-tribal Relations of the Maenge People of New Britain by M. Panoff, July 1969
Bulletin No.31	Inter-ethnic Marriage in New Guinea by A.W. Lind, August 1969
Bulletin No.32	New Guinea Social Science Field Research and Publications, 1962-67 by Susan C. Reeves and May Dudley, October 1969
Bulletin No.33	The Rigo Road: a Study of the Economic Effects of New Road Construction by Marion W. Ward, January 1970
Bulletin No.34	People and Planning in Papua and New Guinea: Papers by D.J. van de Kaa, J.M. Stanhope, T.S. Epstein, N.H. Fry, and C.L. Beltz, April 1970
Bulletin No.35	The Indigenous Role in Business Enterprise: Three Papers from the Third Waigani Seminar, 1969 by A.J. O'Connor, T.S. Epstein and G. Nash, May 1970
Bulletin No.36	Australia and the United Nations: New Guinea Trusteeship Issues from 1946 to 1966 by W.E. Thomasetti, July 1970
Bulletin No.37	Port Moresby Urban Development by J.V. Langmore and N.D. Oram, September 1970
Bulletin No.38	Land Tenure in West Irian: Papers by K.W. Galis, J.V. de Bruyn, J. Pouwer, J.W. Schoorl, and J. Verschueren, December 1970
Bulletin No.39	The Situm and Gobari Ex-servicemen's Settlements by A. Ploeg, January 1971
Bulletin No.40	Land Tenure and Economic Development: Problems and Policies in New Guinea and Kenya: Papers by S. Rowton Simpson, R.L. Hide, A.M. Healy, and J.K. Kinyanjui, March 1971
Bulletin No.41	Would-be Entrepreneurs? A Study of Motivation in New Guinea by Ruth S. Finney, May 1971
Bulletin No.42	Population Growth and Socio-economic Change: Papers from the Second Demography Seminar, Port Moresby 1970, September 1971

Bulletin No.43 Langandrowa and M'buke, Corporate Indigenous Plantations by A.M. McGregor, December 1971

Bulletin No.44 Business and Cargo: Socio-economic Change among the Nasioi of Bougainville by E. Ogan, 1972

Bulletin No.45 Leadership and Socio-economic Change in Sinasina by Sachiko Hatanaka, 1972

Bulletin No.46 Melanesians' Choice: Tadhimboko Participation in the Solomon Islands Cash Economy by I.Q. Lasaqa, 1972

Bulletin No.47 Official and Unofficial Courts: Legal Assumptions and Expectations in a Highlands Community by Marilyn Strathern, 1972

Bulletin No.48 Freight Forwarding in the Australia-Papua New Guinea Trade by P.J. Rimmer, 1972

Bulletin No.49 Hoskins Development: the Role of Oil Palm and Timber - Papers by J.P. Longayroux, T. Fleming, A Ploeg, R.T. Shand and W.F. Straatmans, and W. Jonas, 1973

Bulletin No.50 The Land Titles Commission in Chimbu: an Analysis of Colonial Land Law and Practice, 1933-68, by R. Hide, 1973

Bulletin No.51 Constitutional Development in Papua New Guinea, 1968-73: the Transfer of Executive Power, by P.J. Bayne and H.K. Colebatch, 1973

Bulletin No.52 Chimbu Family Relationships in Port Moresby, by J. Whiteman, 1973

Bulletin No.53 Money, Motivation and Cash Cropping, by T.K. Moulik, 1973

Bulletin No.54 Transition from Subsistence: Cash Crop Development in Papua New Guinea, by R.T. Shand and W. Straatmans, 1974

Bulletin No.55 Matriliny and Modernisation: the Nagovisi of South Bougainville, by Jill Nash, 1974

Bulletin No.56 Foreign Economic Relations of Papua New Guinea, Ross Garnaut (ed.), 1974

Bulletin No.57 Traditional Marriage and the Impact of Christianity on the Solos of Buka Island, by A.H. Sarei, 1974

Bulletin No.58 Co-operatives in Papua New Guinea by Sumer Singh, 1974

Bulletin No.59 Business and Bureaucracy: a study of Papua New Guinean businessmen and the policies of business development in Port Moresby, by C.Leslie Andrews, 1975

Bulletin No.60 Sound-producing Instruments in Traditional Society: a study of esoteric instruments and their role in male-female relations, by K.A.Gourlay, 1975

Bulletin No.61 No Money on our Skins: Hagen Migrants in Port Moresby, by Marilyn Strathern, 1975

Bulletin No.62 Roads and Development in Southwest Bougainville, by Marion W. Ward, 1975

The above bulletins are available at $2.00 each from the A.N.U. Press, The Australian National University, P.O. Box 4, Canberra, A.C.T., 2600, Australia, and the New Guinea Research Unit, The Australian National University, P.O. Box 1238, Boroko, Papua New Guinea.

An annual payment of $10.00 entitles the subscriber to all bulletins issued in the year.